A R G U M E N T:

The Logic Of The Fallacies

John Woods, Douglas Walton

McGRAW-HILL RYERSON LIMITED
Toronto Montreal New York St. Louis San Francisco
Auckland Bogota Guatemala Hamburg
Lisbon London Madrid Mexico New Delhi Panama
Paris San Juan São Paulo Singapore Sydney Tokyo

For Michael Woods and Karen Walton

Acknowledgements

We would like to thank Patrick Suppes for conversations on decision theory; Dick Epstein for discussions and correspondence on relatedness logic; Peter Geach for correspondence on question-asking fallacies; Charles Hamblin and Nicholas Rescher for their published works, which have been very influential in moulding our own approach; Jim Mackenzie for dialectical interchanges by way of the mails; Peter Miller and Vic Shimizu for allowing us to adapt and use teaching materials on the group project exercise in dialectic; John Barker, Nick Griffin, David Sanford and David Hitchcock for verbal or written communications of one sort or another. We are grateful for the interest and support of Jaakko Hintikka, and for his recent contributions to the theory of dialectical reasoning. Also we should like to express our gratitude to Leslie Wilson for compiling the index. Finally, we would like to thank our students over the years who have continually motivated our interest in the fallacies as an object of study.

Preface

This textbook is the first to bring logical theory to bear on the field of informal fallacies and to integrate this field with basic elements of formal logic. The exposition throughout is at the most elementary level, but this book stands apart from the usual texts on informal logic by its attempt to attain a greater depth of analysis of the fallacies. It is our view that the current texts too often acquiesce in the "standard treatment," offering little by way of explanation of the fallacies other than supposedly illustrative examples. Yet, on closer inspection, many of the examples turn out either to be arguments that are not fallacious at all, or arguments in which guidelines are so lacking that a non-arbitrary sorting of the correct from the fallacious cases seems highly problematic or impossible.

This book is designed to offset the prevailing tendency of the "standard treatment" to drift along without clear theoretical guidelines on the fallacies. We begin by looking at the different models of argument that are involved in the various major fallacies, in each case giving enough analysis to let the reader appreciate how the model could be further developed as a theory of argument; the reader will be able to determine what its limitations are. The resulting analyses of the fallacies are subsequently incorporated in the development of an overall cumulative and unifying theoretical structure.

The first model of argument we look at is the quarrel, for quarrelsome dispute is often taken to be the basis of argument. While it is interesting to examine the characteristic types of fallacies that are associated with the quarrel, we find that it is not a model of argument that can take us very far. One problem is that the quarrel is essentially unregulated, and therefore fails to offer reliable guidelines that could be used to settle disputed claims of fallacious argument. Quarrels, as the saying goes, generate more heat than light.

Next we look to the debate, in Chapter Two. The debate is a friend of reason, and in its better forms of expression it can be a noble art. Democratic political and legal systems are built on the principle that the free debate enables us to rationally evaluate arguments. The main problem with the debate as a general model of argument, however, is that it is essentially an adversary procedure: its ultimate goal is to defeat the arguer's opponents. True, debates have objectively agreed upon rules and procedures; thus they are more manageable and orderly than the unruly quarrel. But the primary objective of the debate is to *win the argument*, even if winning involves committing fallacies or using any arguments, however incorrect, to make the judges or referees rule on one's own side. The debate is not regulated in a way to provide a model of argument adequate to evaluating fallacious arguments fairly and objectively. As we show, however, the debate has characteristic fallacies of interest associated with it. Appendix II offers an extended example of a parliamentary debate for analysis, which illustrates both good and not-so-good features of the debate as a model of argument.

Both the debate and the quarrel are subjective, in that they each involve personal interchange – often too personal – between competing parties or groups.

As the next model of argument, we look to an alternative that is purely objective, impersonal and precise – mathematical logic. Perhaps here we can find the exact guidelines and regulatory principles needed to analyse the fallacies. We outline the basics of ordinary propositional logic, which is the basis of all so-called "classical" or standard mathematical logic. We show how propositional logic can and cannot provide procedures to determine correctness or incorrectness of arguments.

We find that classical logic does, indeed, give us the precise and clear guidelines that were lacking in the quarrel and the debate, but at a cost. To yield a simple and reasonable method of testing arguments for correctness, classical logic defines the conditional (if . . . then) in a manner that ignores the question of whether the propositions in the conditional are related to each other. We conclude that, while classical logic is a very valuable tool for evaluating arguments, it is limited by its inability to take *fallacies of relevance* into account. Yet these fallacies are important.

Deductive logic is a valuable model of argument, not least because it provides precisely formulated rules and guidelines. But it seems that there are important arguments and fallacies that elementary classical deductive logic cannot capture or deal with. In Chapters Three through Eight, we look at some of these arguments and fallacies. In Chapter Nine, we see how deductive classical logic can be extended and enriched. We see that one extension of it can take fallacies of relations of topical subject-matters into account. At this point, we will move to a formal semiotic analysis of meaning connections, a system of logical topics.

In Chapter Four we introduce the inductive model of argument, in contrast to the deductive model of Chapter Three. We show that propositions are constructed in a different way in inductive logic than in deductive logic. We survey different fallacies of induction, probability and statistics. We look at a number of common abuses of statistics, including the misleading use of graphs and charts. Finally, we examine different causal fallacies that pertain to induction.

We have two very different models of correct argument: the deductive and the inductive. Thus, before we evaluate any argument, we first have to ask the question, "What kind of argument is it?" The reader learns that there are different levels of analyzing arguments. More is involved than automatically applying a calculus to given data. As Chapters One and Two showed, we have to determine, at the very outset, whether what we are confronted with is indeed an argument. As a minimal requirement, we suggest, an argument should be a set of propositions – one is called the conclusion, the others the premises. Some so-called fallacies do not pass this test of being fallacious arguments – because they are not arguments. But if we *are* confronted with an argument, we have to determine what type of argument it is supposed to be. Only then can the appropriate method of determining correctness or incorrectness be applied.

In Chapter Five, we study yet another model of argument, one that involves an appeal to authority or expertise. It is sometimes a reasonable kind of argument, but one that is open to notorious abuses and shortcomings. We show how the so-called fallacy of appeal to authority is really an umbrella term for a number of specific violations of requirements for a nonfallacious appeal to the authority based on expertise. We see how authority should not be used as a substitute for scientific methods of verification and testing; we see that, in some cases, arguing from authority can be a form of plausible argument. The model of argument involved here – plausible argument – is neither inductive nor deductive, but a weaker type of argument that essentially involves a subjective appeal. Unlike deductive or inductive logic, the plausible model of argument allows us to deal with cases where we are confronted with contradictions. We adopt Nicholas Rescher's method of plausibility screening in order to deal with arguments that begin with an inconsistent set of given data from a collection of authorities.

Up to this point, we are confronted with an apparent pluralism of different models of argument: the quarrel, the debate, deductive argument, inductive argument and plausible argument. Do these disparate models admit of any common conception of argument? Chapter Six provides a general model of argument as a give-and-take, regulated interchange between two or more persons by question and answer sequences. This *dialectical model* of argument is based on the original conception of it given by Aristotle.

As a beginning exercise in dialectic, it is suggested that the student form a discussion group to evaluate arguments on *both* sides of a contentious moral issue. Appendix I gives an illustration of a report based on an actual student project. It shows how to look at both sides of an issue.

As a general model of argument for the fallacies, dialectic can incorporate all the previous models of argument; it can also deal with fallacies that have to do with the asking of questions, as Chapter Six shows. But we see that dialectic requires the good will of the participants to a certain extent. It depends on the willingness of the participants to answer as honestly and informatively as they can, and to ask specific and clear questions. Because of this dependency, it seems not as foolproof in some ways as deductive logic appears to be. However, the advantage of dialectic is that it is much more widely applicable to the fallacies than any of the preceding models. And it can incorporate the other models, including the advantages of precision offered by incorporating deductive logic.

Chapter Seven introduces another model of argument that is clearly not deductive. The fallacy of arguing in a circle involves a seriously incorrect form of argument – but as far as deductive logic is concerned, there is nothing wrong with arguing in a circle. This fact was noted by John Stuart Mill, who even went so far as to propound the scandalous thesis that all deductively correct arguments are instances of the fallacy of arguing in a circle. In this chapter, we examine Mill's claim, and various objections to it, in order to arrive at a perspective on what the fallacy of arguing in a circle amounts to as a logical error. We conclude by showing how dialectic can help us to model circular arguments, and we see how plausibility can be incorporated into dialectic to illustrate what is wrong with arguing in a circle.

To what extent can a theory of the fallacies be a formal theory, given that formal logic seems to be limited in its applicability to fallacies? In Chapter Eight, we clarify the concept of a formal logic by introducing the idea of the *form* of an argument. We look at fallacies that have to do with meaning and grammar, like equivocation and amphiboly, because these fallacies provide testing points for evaluating the extent to which the logic of the fallacies should be formal or informal. Finally, we return to topical connections of meaning as an important aspect of fallacies that remains to be fully analyzed.

Chapters Eight and Nine show how classical deductive logic can be extended to take into account subject matter connections and other aspects of argument of practical importance. First, Chapter Eight shows how classical logic can be constructed as a formal system with precisely formulated rules, requirements and decision procedures. Then it is shown how, by adding a relation to the basic building blocks of classical logic, we can construct an alternative formal system that does take into account connections of subject matters. Here at last is the heralded logic of topical relations. Thus we can see how basic formal systems can be modified and extended to model the concept of argument that is required.

In Chapter Ten, the logic of the fallacies is applied to the practical realities of decision-making in economic reasoning. We see how the fallacies of composition and division are liable to be committed in transitions from microeconomics to macroeconomics and back. These are fallacies that have to do with reasoning about parts and wholes. To study these fallacies, we see that we need to use a formal theory, but one that is richer than classical logic or set theory. Then we look at other principles and rules of economic decision-making, and see that they are dialectical in nature and involve using decision theory as a dialectical tool. The remainder of the chapter introduces the basic concepts of decision theory and shows how these represent models of rationality in economic reasoning.

Chapter Eleven extends deductive logic by introducing the traditional theory of syllogisms – useful in its own way as a practical method of evaluation of deductive arguments – and then showing how modern quantification theory covers the same area with greater generality. Some glimpses ahead into further developments are indicated by introducing basic concepts of modal logic and epistemic logic.

Our basic theme could be adumbrated as follows. When you look at the diverse fallacies, you are struck by (a) how important it is to study each of them as a significant breach of correct argument, and (b) the evident pluralism of different models of argument they embody. We study these different fallacies with an eye to analyzing and understanding the specific type of error that is involved; we also want to construct clear and precise general guidelines for evaluating the argument in order to rule on correctness or incorrectness.

At the present level of theory, an allegation of fallacy is more of a warning than a conclusive indication that an incorrect argument has been advanced, identified and cancelled. If the person accused is able to dispute the point – or if he or she has a defender – the allegation can be too easily dispatched. Merely identifying that the argument seems intuitively wrong by calling it a fallacy, or perhaps by tagging it with a Latin name, is not much use in arriving at some reasonable resolution of the dispute. Reasonable procedures for appealing to or constructing precise guidelines for argument are needed. In short, what we really need is a theory of the fallacies. Putting the elements of that theory into a coherent perspective for the pedagogy of the fallacies is the theme of this book. The book as a whole is a progressive unfolding of an answer to the question, "What is an argument?"

A guide to selected readings is included so that the student can take advantage of the additional stock of examples and specimen arguments that other informal logic texts provide.

Any criticisms, corrections or suggestions for improvements from our readers would be greatly appreciated. Please send them to either of the authors:

Dr. John Woods
Department of Philosophy
University of Lethbridge
Lethbridge
Alberta
Canada
T1K 3M4

Dr. Douglas Walton
Department of Philosophy
University of Winnipeg
Winnipeg
Manitoba
Canada
R3B 2E9

Contents

Argument: The Logic of The Fallacies

Chapter One:
Quarrelsome Arguments

1. Logic: What's It About?

Logic is a principal part of the science of reasoning, and argument is a principal instrument or medium of reasoning. It follows logically that logic involves a study of argument. The main goal of logic is to evaluate arguments, to divide them into two classes, the correct and the incorrect. But, as we will see in this book, there are many different kinds of arguments. A necessary preliminary function of logic is to identify the type of argument one is confronted with in order to see what method of evaluation is applicable. Even prior to this step of identification is the very beginning step: locating the argument itself. The first question to be asked is: what is the argument?

The core of an argument is a set of propositions that may be divided into two classes, the *premisses* and the *conclusion*. The premisses are the basis of the argument; the conclusion is what the premisses are being used to argue *for*. A third part of any argument is the *evidential link*, which relates the premisses to the conclusion. Much that surrounds these three basic elements is also part of the argument. But if there is no core – propositions joined by an evidential link – then there is no argument. It is a necessary condition of argument that there be a set of propositions, the premisses and the conclusion of the argument.

An argument is a set of propositions, the premisses and the conclusion.

What is a proposition? A proposition is something that is true or false. We may not know, in fact, whether it is true or false, but if it is a proposition it must be either true or false. Questions and commands are not propositions, but they may contain propositions. A proposition is that which is asserted by a declarative sentence in any language. A proposition is not a purely linguistic entity. The sentences, "Snow is

1

white" and "La neige est blanche," are quite different sentences, but they may express the same proposition.

Exercise

Determine which of the following are propositions.
1. The cat is on the mat.
2. The cat was on the mat.
3. The cat will be on the mat.
4. I believe the cat is on the mat.
5. Either the cat is on the mat, or he isn't.
6. He is tall and bald.
7. Please wait your turn.
8. What's it all about?

Given a set of propositions that supposedly make up an argument, how are we to tell which is the conclusion? Answer: there is usually a key word or phrase ("therefore," "consequently" or "we may conclude that") which indicates that one proposition is being selected as the conclusion. However, there are many different ways of indicating that a conclusion is being put forward on the basis of argument. Some of these ways are subtle; any would-be critic of an argument should take care to see that he has correctly chosen the conclusion.

It is not always clear which part of an argument is meant to be the conclusion and which propositions are meant to be the premisses. As we will see later, many a fallacy occurs when conclusion and premiss are mixed up. A fallacy is an argument that is a tricky deception because it is incorrect even while it has a tendency to seem to be correct. The unwary, who may not be acquainted with how the fallacies work, are especially vulnerable. Our principal task in this book will be to study the fallacies.

Now we know that arguments are made up, at least in part, by propositions, but we still do not have a very full grasp of what an argument is. What is an argument like, as a whole? All of us are familiar with arguments; we are confronted with them every day, in advertising, school work, scientific pursuits, ethical discussions, even in deciding what to do each day and deliberating about the alternatives. Probably most of us associate an argument with a contentious dispute where claims are advanced and subjected to justification or refutation. Certainly one type of argument is the quarrel or contentious dispute.

2. Quarrels

Two kinds of quarrels are familiar.

(1) The "Yes, you did /No, I didn't" quarrel. Sue insists that Bill told her he would spend last Friday in the library. Bill, who was spotted at

the races, contends that he had told Sue he planned to relax at the races. The disagreement is basically about *what the facts are*. As anyone who has ever had such a disagreement knows, it can be abusive and inconclusive. Admittedly, sometimes such arguments are great fun; yet more often they are a nuisance for, or an impediment to, reasoning.

(2)   The "You don't love me any more" quarrel. Bill is racked by doubts about Sue's behaviour. When they were first in love, Sue seemed to hang on Bill's every word. She laughed at his jokes, even those that made his pals groan. She reserved all her free time for him. Recently, however, Sue hasn't hung on Bill's every word, and once she drifted off to sleep when Bill was regaling the others with tales of his adventures as a moose hunter. Not only does Sue no longer laugh at Bill's every joke: yesterday she remarked that the one about the farmer's daughter and the astrophysicist was "mindless and sexist." And last week, she went to a concert with John.

> *Bill:* You don't love me.
> *Sue:* But I do! How could I call you a silly ass without caring for you enough to risk giving offence, perhaps even losing you?
> *Bill:* Sure, sure. How do I know that this isn't just what you have in mind – losing me, as you call it, or breaking up, as I would say?
> *Sue:* Bill, you just don't understand. Be reasonable.

Here, then, the disagreement is not so much about what the facts are, but about what the facts mean, how they are to be interpreted, about *what the facts imply*.

Using our logical terminology, we may say that the part of an argument dealing with what the facts are is the *premiss of the argument*; the part dealing with what the facts imply we call the *conclusion*.[1]

A logician would say that, in the first quarrel, Sue and Bill disagree about premises; in the second quarrel, they disagree about the conclusion. In the first quarrel, we find *premissory instability*: that is, we find that Bill and Sue agree about very few, if any, of the salient facts of the case. They have few or no premises in common. In the second quarrel, we find *conclusional instability*: that is, we find that Bill and Sue, though agreeing about most of the salient facts, cannot agree about what the facts imply; thus they cannot agree about what should be *concluded* from the facts.

It is not hard to see why instability poses difficulties in the evaluation of an argument. If there is both premissory and conclusional instability in the arguments of Bill and Sue, then we have *two arguments*, which do not interact with each other at all. If we try to locate the premises and the conclusion, the logical core of the argument, we can't find them. Bill has his premises and conclusion. Sue has her premises and conclusion.

But we can say nothing about the argument as a whole. Where there is no common ground, there is no argument; instead, there is a multiplicity or plurality of unconnected arguments. Small wonder, then, that the quarrel tends to be neither productive nor enlightening from a logical point of view.

We can now offer the following helpful generalization about the quarrelsome argument:

An argument is a quarrel if it suffers from premissory or conclusional instability, or both.

It is easy to see why quarrels are such noisy, personal and inconclusive affairs. If, as in the case of premissory instability, we cannot even get started on the road to agreement, then frustration, accusation and hurt feelings are bound to occur. References will tend to become personal and disagreeable. Sue might eventually complain that if Bill can't recall what he said last Friday, then he is a simpleton; Bill might retort that Sue is a hysterical shrew. Before you know it, things will have taken another nasty turn.

Similarly, having got the discussion nicely under way with some basic premissory agreement, things might come grinding to a halt owing to a lack of common conclusions. Then the same personal disruptions could occur. Bill might contend that Sue shows herself to be a "typical woman" in having no capacity to reason beyond her nose, or to perceive what follows from what. And Sue may earnestly offer to slap Bill's moronic face (and perhaps be forgiven for it).

Why, then, do contenders in such arguments become quarrelsome? With no prospect of getting started and no prospect of reaching a conclusion, they have nothing left to do but fight. The moral would seem to be:

Premissory instability and conclusional instability cause *frustration*; frustration causes *aggression*.

---

3. The Functions of Logic

As we have seen in the past few pages, one function of logic is to offer a *description of the structure of various kinds of argument*. For example, an argument can be said to be made up of premises and a conclusion; if the argument in question is also a quarrel or a spat, it can be said that it exhibits a fairly high degree of either premissory or conclusional instability or both.

Premissory and conclusional instability can, in their turn, be described as radical disagreement among the participants about which premises are *true* and which are *false*; or about what does and what does not follow from these premisses. However, in addition to describing the logical structure of arguments, it is also a function of logic to *evaluate* their *success* or *failure*.

The success or failure of an argument depends at least partly on what it was intended to accomplish. For example:

1. it may have been intended as a completely knock-down *proof* of its conclusion; or
2. it may have only been intended to establish that its conclusion has a *fairly high probability*, or that it is more likely than not;
3. it may have also been intended as a *refutation of one or more of its premisses*. For example, sometimes when we argue we assume an opponent's premiss "for argument's sake"; then we show that something plainly false follows from that premiss. By doing this, we also show that our opponent's premiss is false. (It is a basic rule of logic that propositions implying falsehoods are themselves false.)

An argument may fail of its intended purpose and yet succeed in some other way, perhaps unintentionally. For example:

4. I might intend to use my argument as a knock-down proof, and yet it might only show the conclusion to be a fairly reasonable and credible one. Consequently, my argument may be a bad *proof* but nevertheless a good *argument*.

Thus in evaluating an argument, we must keep in mind both intended and unintended success or failure.

It is useful to remember that logic is the science of reasoning and that reasoning is vitally and essentially concerned with the orderly improvement of our understanding of things. So, among the natural allies of the concept of reasoning, we may expect to find the following notions:

Evidence
Justification
Warranted Inference

These concepts would seem to interrelate in this way: one has done a *good* job of *reasoning* when one's *evidence* is strong enough to *justify* one's conclusion, and to make one's drawing of that conclusion a *warranted inference*.

But argument frequently accomplishes more than to come up with a warranted inference based on justifying evidence. For example, an argument's purpose is sometimes *persuasion at any price*, or even *out and out deception*.

There are, then, two basic kinds of evaluative interest that logic serves in arguments. One is *approving and positive:* it explores and recommends the various ways and means in which arguments can serve the proper objectives of competent and responsible reasoning. The other is *disapproving and negative:* it explores and cautions against deceptions, snares and delusions, fallacies that can deter and frustrate the proper object of competent and responsible reasoning.

The vocabulary of logic has provided us with a description of the structure of arguments, and we know how to describe the structure of a quarrelsome argument. But we might also wish to know how to evaluate such arguments. What, then, is the logical evaluation of a quarrel?

4. Quarrels: Their Logical Evaluation

By and large, logic evaluates the quarrel negatively and disapprovingly. True, a good fight may do wonders for a person's spleen, perk up a person's love life or test the combatant's mettle, but fights only rarely and unintentionally advance the cause of reason, and usually they are a serious impediment to one's understanding of things. The logical lessons to be learned are therefore mainly negative:

1. A quarrelsome argument has little or no chance of starting or concluding satisfactorily. *Consequently, quarreling is a systematically inefficient and inconclusive mode of reasoning.*
2. Premissory or conclusional instability in an argument makes that argument a potential quarrel. Therefore, *it is not reasonable to proceed with such arguments until their instability is either eliminated or reduced.*[2] If an argument's instability shows no promise of being reduced, one should stop talking, agree to disagree or change the subject.
3. Premissory or conclusional instability make for frustration, and frustration makes for aggression. Aggression is a well-known sponsor of fallacious reasoning. Therefore, there is a second major reason to disentangle oneself from an argument that threatens to be a quarrel. Not only is the argument likely to be inconclusive; as well, *the argument risks the use of fallacious reasoning.*

5. The Fallacies: A Preliminary Word

This book is chiefly concerned with the logical description and evaluation of the *fallacies*. A fallacy is a pitfall of reasoning that exhibits a general and recurring tendency to deceive and to deceive successfully, to trick even the entirely serious and honest arguer. We could say of the quarrel that it represents the *anarchy of argument*; we could say of the fallacy that it represents the *deceit of argument*. It is in their capacity to deceive, to trick and to obscure that fallacies can be the ruination of reasoning. And yet how easy it is for us to fall into the clutches of the fallacies; we can be duped by them and we can also commit them.

It would be well to have an indication of the kinds of fallacies that quarrels characteristically give rise to. We shall confine ourselves to two important fallacies in this chapter.

6. The Fallacy of the Ad Baculum (or Appealing to Force)

In the scenario a few pages back, we imagined Sue and Bill in the heat of battle, and Sue threatened to slap Bill's face. This is not a very impressive example of the fallacy of appealing to force because it is too obvious to fool anyone. However, it takes little imagination to conceive of a situation in which an arguer repeatedly resorts to innuendo, yet in which his use of innuendo is so subtle that it goes unnoticed. In fact, this makes resorting to it worse still. The innuendo is not detected, and it also can create the eerie and even frightening impression that I am doing myself no good whatsoever in sticking to my side of the argument, and that *somehow* I am placing myself in jeopardy.

Item of Interest. During his visit to the UN General Assembly, USSR Premier Nikita Kruschev banged his shoe on a table and chided the representatives of the Western nations with the remark, "We [the Soviets] will bury you." Most commentators pooh-poohed this as a perfectly obvious and gross appeal to force, a threat; and they dismissed the incident for its ineptness and heavy-handedness. In fact, however, Kruschev's intentions were much more subtle. He was actually referring to the doctrine of Marxism-Leninism that teaches that capitalist countries, by necessity and because of their own inherent contradictions, will be destroyed and eliminated by their own unavoidable dysfunctions. So Kruschev was mocking the West, jeering at its quite hopeless efforts to oppose Soviet expansion, or to survive. In effect, Kruschev was saying, "Why bother? Your cause can't be the right one, for history itself is going to extinguish you."

Like many a fallacious stratagem of rhetoric, appeals to force are effective because they appeal to strong emotions. Such appeals are effective not so much as an argument but as a distraction from argument. Aristotle pointed out cannily that the strategy of simulating anger and prompting one's opponents to contentiousness is a highly effective tactic.[3] If your opponent seems to be winning the argument by virtue of the logic of his propositions and proofs, and you yourself are at a loss to find logical evidence for your own position, an all-too-effective strategy is to derail the argument by unleashing powerful emotions. Aristotle observes that arguers are less able to defend their argument when emotionally agitated. One way to produce anger is to make it clear to your opponent that you want to use unfair means in the argument; indicate subtly that you might even resort to violence or force. This is the *ad baculum*.

A not-so-subtle example of the *ad baculum* is furnished by R. Grunberger, author of *A Social History of the Third Reich*. He reports that the Nazis used to send this reminder to those who had let their subscription to the Nazi newspaper lapse: "Our paper certainly deserves the support of every German. We shall continue to forward copies of it to you, and hope that you will not want to expose yourself to unfortunate consequences in the case of cancellation."

We said, at the beginning of this chapter, that a primary task of logic is to locate the premises and conclusion of an argument. The *ad baculum* is used to subvert this objective: the goal of the *ad baculum* is to make us lose track of the argument altogether in the heat of emotion.

It is important to recognize that the *ad baculum*, although it is called a fallacy, need not be an incorrect argument. Indeed, the *ad baculum* is probably not even an argument at all. It is, rather, an attempt to cover up or nullify argument. If I point a revolver at your temple to win an argument, what premiss do I advance? If I threaten, "Accept this conclusion or I'll shoot!", what proposition do I enunciate? Have I stated a proposition at all? Remember that a proposition is something that is true or false; what we have here seems more like an imperative or command. In general, it is hard to see how a threat of force or violence can be a proposition or a set of propositions. And it is hard to see how the *ad baculum* is an argument. It would seem, therefore, that to resort to an *ad baculum* is to abandon argument altogether, or at least to move away from argument.

We will see later that generally a fallacy is a special type of incorrect argument. Some stratagems of argument, like the *ad baculum*, are not "fallacies" in this sense, because they do not appear to be arguments at all. But the *ad baculum* is called a fallacy, and we can easily appreciate its usefulness to those who would subvert argument. Those who use such a ploy do not advance an incorrect argument; instead, they obsure the argument to the point where the premises and conclusion have been lost track of, forgotten or submerged into extraneous distractions.

Exercise

Determine whether an *ad baculum* has occurred in the following arguments.

1. An interviewer, wishing to play the rôle of the unconvinced skeptic, asks a fundamentalist preacher what reasons he can give for believing in an afterlife. The preacher replies: "Do you want to go to hell? There are two possibilities – eternal salvation or eternal suffering. If you accept Jesus you will

avoid going to hell. You have the choice of whether to believe or not. But if you don't believe, you will go to hell."

2. A famous movie actress doesn't get along with her leading man; as a result, he is fired from the cast. The reason cited by the actress is "poor chemistry" between the two. The actor retorts, "We're actors, not pharmacists." The actress replies, "The manager was the one who spoke to him. If I find out he has put the blame on me there will be one manager less."

3. According to a Reuter article, the Ayatollah Khomeini, head of state in Iran, spoke to a delegation of air force officials as follows: "Our government is Islamic and you must support the government of theologians so that no harm can be done to the country. You who are not well versed in Islam, do not get in the way. The nation has voted for the Islamic republic and everyone should obey. If you do not obey, you will be destroyed."[4]

4. A National Front political leader in England, defending his party's dire warnings that racial minorities will one day become majorities through population increase and then subject whites to harsh treatment: "What's wrong with appealing to fear? Lots of people are afraid. Therefore it's a legitimate political factor to respond to it. There's nothing immoral about it."

7. The Fallacy of the Ad Hominem (or Speaking of the Man, Not His Argument)

In the scenario involving Bill and Sue, there are at least two occasions when the contenders resort to personal abuse. Bill stands accused of being a simpleton, and Sue of being a typical woman, who therefore is no good at reasoning. The abuse is obvious, and so obviously misconceived or irrelevant as to stand little chance of success as a deception. (Though if passions run high enough, even a gross and stupid remark can be enraging and can, therefore, subvert the objectives of the argument.)

In a more harmful form, the *ad hominem* argument is not abusive but, as logicians sometimes say, *circumstantial*. An opponent commits the fallacy of the circumstantial *ad hominem* when he refers not to his rival's argument but rather to some fact or alleged fact about the man himself. (Hence the phrase "ad hominem," which is Latin for "to the man.")

The cleverest circumstantial *ad hominem* is one in which reference to one's opponent is perfectly true, and well-known to be perfectly true. Here is how the stratagem works: if the remark is true, then how could such a reference be contrary to the purposes of competent and responsible reasoning? For is not truth the natural ally of reasoning? Well, consider this.

Bill and Sue are battling, as usual. The point of contention is whether abortion on demand is justifiable. Sue says in utter exasperation, "Blast you, you're a man, you just can't see it!" True, Bill is a man. Obviously. But the potential for damage lies not in what Sue is saying, true as it is, but rather in what she leaves unsaid and only implied. Her implication is that a man will oppose abortion on demand because, as a man, he cannot experience unwanted pregnancy, and accordingly cannot help himself in opposing abortion. Because Sue's accusations are not explicitly stated, they might have an *undetected* effect on the argument.

Item of Interest: This example crops up repeatedly in contemporary discussions about abortion. Many people commit the "you are a man (so you can't help being opposed)" fallacy. Ironically, they seem not to be aware that women are open to a similar attack. Women, presumably, are in a position to experience an unwanted pregnancy; thus, presumably, they are unable to help themselves in favouring abortion on demand. If you think that neither you nor your opponent can help what side you are on, then what in the world is the use of continuing the argument?

The moral is that even when the truth is told in an *ad hominem* fallacy (and sometimes it isn't), it is *irrelevant to the point of argument*. Some further examples will illustrate.

The circumstantial *ad hominem* arises when there is a logical clash or opposition between the thesis that some participant in argument is putting forward and some circumstances pertaining to that participant himself. For example, a person might recommend some policy as a worthwhile principle, but the circumstances indicate that he has not followed – or is not following – that policy himself. Any person caught in such an apparent conflict can be highly vulnerable if his opponent attacks *ad hominem:* "You don't even practise what you preach!" He seems caught in the most ridiculous kind of illogicality. He tells us we ought to do something, but he doesn't even do it himself. The *ad hominem* can be a crushing rejoinder.

But be careful not to overestimate its force. A person's inability or failure to follow the advice embodied in his own thesis is interesting, significant, and even comical, but remember that his thesis may be based on correct arguments. A speaker from Alcoholics Anonymous may be giving us good advice on how to avoid becoming an alcoholic, even though he himself was an alcoholic at one time. True, he did not once practise what he now preaches, but that makes his first-person advice all the more valuable. Alternatively, there is no reason to reject it as worthless. *Ad hominem* is tricky, and we should not reject an argument too quickly when we detect a circumstantial inconsistency present.

On the other hand, some circumstantial inconsistencies are more significant than others. If our speaker from Alcoholics Anonymous showed signs of inebriation while giving his speech, we might have legitimate doubts about his capability as a spokesman on this particular subject. Another example of a significant circumstantial inconsistency: a Citizen's Committee protests the Government's proposal to substantially raise the salaries of Members of Parliament, on the ground that the Government has recently entreated the Canadian public to resist such inflationary practices as wage-escalation. The Citizen's Committee has a good point. Noting circumstantial inconsistency is a legitimate move in argument. Even so, it would be too hasty to conclude that the Government's position is totally wrong or that their advice is silly. The fallacy occurs where the *ad hominem* is carried too far.

A characteristic climate for the *ad hominem* is the following sort of dialogue. *Parent* has just finished citing evidence of links between smoking and lung cancer, and other evidence of bad effects of smoking on health.

> *Parent:* So you see, smoking is a bad business to start.
> Once you start, it is difficult to stop.
> *Child:* Yes, but you smoke. How can you tell me not
> to smoke if you do it yourself?

Child has a worthwhile point. If Parent is not following the course that he himself recommends, could this be evidence of insincerity? But Parent's arguments that smoking is bad for health may be based on good evidence. Child should not reject Parent's argument solely because of the circumstantial inconsistency. Two questions need to be answered. (1) Is Parent's thesis substantiated by the evidence he cites? (2) Does Parent's recommendation conform to his own circumstances?

What is the fallacy of the circumstantial *ad hominem*, then? It is making the unwarranted leap from the premiss of circumstantial inconsistency (the claim that the arguer's circumstances clash with his thesis) to the conclusion that the argument as a whole is worthless (the claim that the arguer's thesis must be false).

8. Another Form of the Circumstantial Ad Hominem

As we have seen, a circumstantial *ad hominem* argument has as its basic core a set of two propositions that are inconsistent with one another. One proposition is a thesis recommended by the arguer; the other proposition describes some circumstance of the arguer himself. These two propositions may be described as the basic premises of the *ad hominem*. But sometimes a fallacy occurs in an *ad hominem* argument when the very premises are incorrectly described or confused. This is simply a failure to correctly locate and identify the argument in the first place. What can happen is that a third proposition, which resembles one of the original pair, is illicitly substituted for one of the original premises in the argument. When this happens, there may appear to be an inconsistency where none really exists. The result is a subtle form of *ad hominem* fallacy.

An interesting study case is the notorious *hunters' argument* reported by Archbishop Whately. The argument goes something like this.[5]

> *Critic:* How can you derive pleasure from gunning down a helpless animal? Surely the killing of deer or trout for amusement is barbarous.
>
> *Sportsman:* If you're so concerned, why do you feed on the flesh of animals? Aren't you being inconsistent?

What Sportsman is suggesting is that Critic's actions are inconsistent with his moral pronouncements. Sportsman rebuts Critic by using the *ad hominem*.

But where is the inconsistency? Well, according to Sportsman, Critic eats meat but at the same time asserts that hunting is morally repugnant. Is this inconsistent? The answer is no. It would be inconsistent to hunt while asserting that hunting is morally repugnant. But there is no inconsistency in condemning butchery for amusement while eating a steak dinner. There is a difference between hunting and meat-eating.

Augustus De Morgan explained the fallacy as an illicit substitution producing what only appears to be an inconsistency. "... It is not absolutely the same argument which is turned against the proposer, but one which is asserted to be like it, or parallel to it. But *parallel cases* are dangerous things, liable to be parallel in immaterial points, and divergent in material ones."[6] We can more clearly see the illicit substitution if we contrast the genuinely inconsistent pair [X asserts that hunting should not be brought about; X brings about hunting] with the parallel pair [X asserts that hunting should not be brought about; X brings it about that meat-eating obtains]. The parallel pair may superficially look like the inconsistent pair, but there is really a big difference. Not perceiving this difference gives rise to the fallacy.

Although the sportsman accuses the critic of a circumstantial *ad hominem*, in this case the roles are reversed: it is really the sportsman who commits an *ad hominem* fallacy. His accusation of circumstantial inconsistency is unfair and fallacious.

In an argument like this, how can we tell who has committed the *ad hominem* fallacy? Many textbooks say that whoever intended to deceive his opponent committed the fallacy. But according to our analysis, it is the argument itself that is or is not fallacious, not the participant or his intentions. We reiterate that an argument is a set of propositions. The fallacy in this type of circumstantial *ad hominem* is contained in the relation between the two propositions themselves: they are not inconsistent with one another. Sportsman's argument is fallacious not because he claimed that Critic was inconsistent, but because of the way he stated the *ad hominem*, the inconsistency did not really obtain. Sportsman simply got his premises wrong.

We will learn more about inconsistency as this book proceeds. We will see that logical inconsistency occurs where a proposition is asserted and negated at the same time. Thus to assert that "the cat is on the mat, and the cat is not on the mat" is to assert a proposition that is logically inconsistent. In examining the *ad hominem*, we have not been discussing a logical inconsistency, strictly speaking; instead, we have seen the inconsistency between what one *causes to happen* and what one *says should happen*. To go back to our example of the cat, what I assert is inconsistent in this sense if I remove the cat from the mat while saying that the cat should be on the mat. We call this kind of inconsistency "deontic-praxiological inconsistency," which means that it has to do with norms and actions respectively.[7]

A speaker can be involved in a circumstance, or even create a circumstance himself, which not only makes what he says inconsistent with that circumstance, but also actually *refutes* what he says. A very simple example: Bill argues, "Since I never speak English, what I am now saying is not in English." It is worth noticing that a correct attribution of the circumstantial *ad hominem* can be conclusion-defeating: not all uses of the *ad hominem* are incorrect or fallacious.

If a pair of propositions is logically inconsistent, one of them must be false. But it does not follow that further argument cannot remove or explain the inconsistency. An *ad hominem* can be defused. After the Second World War, Bertrand Russell advocated attacking the Soviet Union as a possible response to failure of negotiation; then, in the fifties, he advocated the "Better Red than dead" argument. Is this inconsistent? Yes, but Russell explained why he had changed his mind. Before the Russians had nuclear weapons, Russell thought that aggressiveness was a rational policy; but such a policy would be irrational in dealing with a country that had nuclear capability. In short, Russell gave good reasons for his change of mind. Therefore, given the overall logic of Russell's position, there is a reasonable basis for defending him against an allegation of *ad hominem*.

Exercise

1. Tommy Douglas wrote that once when he was bemoaning the plight of Western agriculture in the Commons, some Liberal members from Saskatchewan called out, "What do you know

about it? You've never farmed." Douglas answered, "And I've never laid an egg either, but I know more about omelets than most hens."[8] Did anyone commit a fallacy here?

2. In the Manitoba legislature, someone in the Conservative Party is accused by an NDP critic of misusing funds to support a questionable youth-grant project. The reply to the critic was that he should be careful in pressing this sort of criticism, given the notorious record of his party, when it was in power, of squandering money on youth grants. Has anyone committed a fallacy?

3. We are told by Boswell that Dr. Johnson asked someone who argued for the equality of all mankind – in the presence of a servant – would the arguer allow a footman to sit down beside her? Another time Dr. Johnson is said to have remarked, after listening to a man who had cleverly argued that the difference between virtue and vice is illusory: "When he leaves our house, let us count our spoons." Could Dr. Johnson be fairly accused of committing a circumstantial *ad hominem* in either of these remarks?

4. Evaluate each circumstantial *ad hominem* argument that occurs in the following article.

There are those who claim that the fictional 007 became a worldwide glandular and intelligence hero because John Kennedy said that he was a James Bond fan.

It is established history that after Lyndon Johnson had the Fort Worth barbecue wizard Walter Jetton at the White House, Jetton's vans, filled with succulent ribs, were summoned by hostesses all across the land.

When Richard Nixon pinned a flag in his lapel and became the spirit of '76, lapel flags blossomed in board rooms and Rotarian halls. After the story got out that Nixon had seen *Patton* at least three times, the motion picture's gate went up an estimated 20%.

One of the immense powers of the presidency is the power of personal suggestion and example. In fact, television has so greatly magnified the human elements of Presidents that this may be as important to White House leadership as the constitutional authority of the office. For better or worse, television has made the President "somebody very close" to most American citizens, says Pollster Daniel Yankelovich, and while their own feelings of inadequacy and humility keep them from making instant judgments about complicated issues like milk price supports and the Middle East oil tangle, Americans seize on the personal actions that they can see in their living rooms and can understand. History may prove that Nixon's worst failure is the sequence of seemingly minor personal absurdities that he indulged in.

Resentment of the President's inconsistencies is now deep in the American soul. Nixon preached law-and-order but presided over a lawless administration. While he was cutting programs of education and health and urging personal spending restraint on everyone else, his private homes were being voluptuously appointed at taxpayers' expense. His calls for all Americans to carry the national commitments were still ringing when it was learned Nixon had used gimmicks to reduce his taxes to a pittance. And even as he belatedly began to recognize the seriousness of the energy crisis, he roared round the country in his huge jet and churned up and down the Potomac valley in his big helicopters.

While Nixon has decried distortions in the press, his own arguments have been accented with inaccurate historical allusions and downright misstatements that he has never bothered to correct. Cropping up now as a public worry in the opinion samplings is another of those "petty" episodes that the men in the White House swat as if they were mere flies. Nixon went into a meeting with 16 Governors and told them he knew of no other Watergate developments that would embarrass them. The next day it was revealed that one of the tapes had a more than 18-minute gap and Nixon had known about it. The man who is "somebody very close" had deceived not just 16 Governors but also millions and millions of his people.

The grimly comic sequence of how the long buzz got in the tape is now registering on the public mind. A vast number of Americans know a good deal about tape recorders, and they can follow the electronic saga. The final fragments of credibility in the tapes were shattered in many minds.

One can predict with some confidence that yet more disapproval of Nixon will come out of the White House's cannibalism. Not only have Nixon and his few confidants turned a cold shoulder on many of the young presidential aides caught in Watergate, but they have also tried to smear the reputation of former Attorney General Elliot Richardson and are now discrediting White House Counsel Fred Buzhardt. If Buzhardt devised the ludicrous Watergate legal strategy, he deserves criticism. But publicly humiliating a loyalist like Buzhardt is another of those small human rituals that most people comprehend.

What stands now between Nixon and impeachment, suggest some of the opinion diagnosticians, is a thin tissue of personal well-being felt by most Americans. They still have it pretty good, and they don't want a change. But if too many of them lose their jobs or their mobility or their heat, then their fear and disillusion may be turned with even greater force on the man they see so often in their living rooms, who has disappointed them in so many personal ways.[9]

Notes

1 In this book we adopt the spellings "premiss" and "premisses." The American philosopher C.S. Peirce points out that the spelling appropriate to logic derives from the medieval Latin *praemissa*. The spellings "premise" and "premises" (as in "He was removed from the premises for trespassing") come from the French *premise*, as in the expression *les choses premises*, used in inventories. We came upon this point in T.A. Goudge's influential book, *The Thought of C.S. Peirce* (Toronto: The University of Toronto Press, 1950, page 5).

2 Or quarantined by procedures used in debates and criminal trials. These procedures are discussed later.

3 Aristotle, *De Sophisticus Elenchis*   174a.16.

4 The article was reprinted in the *Winnipeg Free Press*, 20 September 1979, page 16.

5 Archbishop Whately, *Elements of Logic* (New York: William Jackson, 1836), page 197.

6 Augustus De Morgan, *Formal Logic* (London: Taylor and Walton, 1847), page 265.

7 John Woods and Douglas Walton, "Ad Hominem," *The Philosophical Forum* 8(1977): 1–20.

8 T.C. Douglas, "Reflections of a Hard-Headed Dreamer," *Reader's Digest* (April 1979): 95.

9 Hugh Sidey, "Failings of Somebody Very Close," *Time* (10 December 1973): 19. Reprinted by permission from TIME, The Weekly Newsmagazine; Copyright Time, Inc. 1973.

Chapter Two
The Debate

Another type of argument we are familiar with is the debate. The debate appears to have more structure and orderliness than the quarrel, so perhaps we might look to the debate as a model of argument that is useful in approaching the study of fallacies. In a debate, there are winners and losers, and definite rules determine the outcome. Some guidelines might be derived from the debate to enable us to evaluate and identify arguments in science, ethical disputes and other everyday transactions and activities where arguments occur.

Like the quarrelsome argument, the object of a debate can be to frustrate the rightful rôle of reason; like the quarrel, the debate can be a noisy, personal and fractious affair. In fact, a failed debate may quickly deteriorate into just another quarrel. But at its best, a debate can be a noble thing – if not reason's handmaiden, then perhaps reason's court jester – making obvious the ridiculous, challenging the dogmatic and putting to the critical test even a cherished principle or postulate.

1. The Debate: Its Description

A quarrel, as we said, is the *anarchy of argument*. Debates are *rule-governed* enterprises, presided over by a referee who is bound to fairness and objectivity.

A debate is not meant to end in agreement between participants; it is settled not by the debaters but by a panel of judges (or by a vote of Parliament or of Congress); even then, the agreement need not be unanimous. This is an extremely important (and fortunate) feature of debates, and is well worth emphasizing.

In a debate, total agreement among the contending parties or the panel of judges or the body of voters is not necessary. A simple majority from among the judges or the voters will do.

Thus debates can tolerate a very high degree of premissory and conclusional instability among the contending participants. Unlike quarrels, where the object is to reach a common agreement between the participants and where that object is frustrated, debates are not liable to the anarchy and chaos of instability.

It is important to be clear about this point. Debates can tolerate *total premissory or conclusional instability* between the contending participants or debaters. However, debates always involve a non-contending participant, namely, that person or body or group that is the judge. Obviously, then, if a debater is to be successful, he needs a fairly high degree of premissory and conclusional stability between himself and a simple majority of the judges. We assume that debates are relatively business-like contentions, designed to produce *bona fide* decisions. Sometimes, of course, a debate is arranged merely for fun, and is won or lost on funniness or wit or personal devastation; in such a debate, it is not important if the judges or the audience agree with the substance of what is actually said. This kind of debate (though we don't deny that it sometimes occurs in our legislatures!) is a limiting kind of argument; it provides, for the logician, only negative lessons.

One of the virtues of a debate is that it is *refereed*, and so is governed by

> rules of procedures
> rules of conduct
> rules of decision

Debates, then, are exchanges that are meant to be fair, and the main benefit of such a system is that anarchy is averted. Nevertheless, it is also important to recognize that this very "benefit" of debates gives rise to special corruptions all of its own. For bear in mind that a referee system is a kind of external authority; it inhibits, prohibits or prevents a debater from having things just his own way. Thus there is the temptation to evade the weight of authority not by anarchy, but by

> sophistry
> insincerity
> flattery
> ambiguity

The debater is tempted to say something he clearly does not mean; what he says stays within the rules, but what he *means* does not.

The logical structure of a debate is a fairly straightforward matter in its essentials. A debate is:

a contest for approval

a contest between two or more contending participants (they may be individuals or groups of individuals) a contest that is presided over by a referee or a speaker, whose function it is to maintain order, to apply and interpret the rules and, ultimately, to bring the debate to a vote

a contest in which the verdict is usually by a majority vote, either of a panel of judges or a jury or members of "the House". The winning side thus wins "the approval of the House".

The first temptation of debate that we noted is sophistry. What is sophistry? A clue is found in Aristotle. In his early writings, a sophism is simply the conclusion of what Aristotle called a "contentious" argument. By a contentious argument Aristotle meant an argument based on beliefs generally accepted but nevertheless untrue, or an argument in which there only seemed to be proper reasoning for such beliefs. In some of Aristotle's other writings, in the literature of later periods and in this book, "sophism" refers, quite generally, to any fallacy.

So sophistry could be described as the use of tricky or subtle fallacies to confuse or deceive an opponent. More simply, sophistry is the use of fallacies in argument.

A good example of sophistry is found in Plato's *Euthydemus*. In this dialogue, Euthydemus and Dionysodorus are sophists who turn their attention to a young chap, Clinias, and devastate him with their trickery.

> "But now," said [Dionysodoros], "is Clinias wise or not?"
> "He [Clinias] says, not yet . . . he's no boaster, you know".
> "And you people," said Dionysodoros, "want him to become wise and not to be a dunce?"
> We agreed.
> "Then you wish to become one that he is not, and no longer to be one that he is. . . . Since you want him no longer to be one that he is now, you want him to be destroyed, it seems!"

The second temptation in a debate is insincerity; the third is flattery. The following excerpt contains examples of both of these temptations. The passage is Mark Anthony's speech from Act 3, Scene 2 of Shakespeare's *Julius Caesar*. It should be borne in mind that Brutus, to whom Mark Anthony repeatedly refers, has himself just addressed the same throng and openly admitted his part in the murder of Caesar.

> Friends, Romans, Countrymen, lend me your ears;
> I come to bury Caesar, not to praise him.
> The evil that men do lives after them;
> the good is oft interred with their bones;
> so let it be with Caesar. The noble
> Brutus hath told you Caesar was ambitious;
> if it were so, it was a grievous fault;
> and grievously hath Caesar answer'd it.

> Here, under leave of Brutus and the rest, –
> for Brutus is an honourable man; so are they all,
> all honourable men, –
> . . . come I to speak in Caesar's funeral.
> He was my friend, faithful and just to me:
> but Brutus says that he was ambitious;
> and Brutus is an honourable man.
> He hath brought many captives home to Rome,
> whose ransoms did the general coffers fill:
> did this in Caesar seem ambitious?
> when that the poor have cried Caesar hath wept;
> ambition should be made of sterner stuff:
> yet Brutus says he was ambitious;
> and Brutus is an honourable man.

Here is another example of insincerity. We see that the rules of procedure of the Parliamentary debate are openly violated by both participants and that the wrongdoers merely feign their apologies. This exchange has been attributed to at least four different pairs of English Parliamentarians from the time of Henry VIII onwards.

> *First Parliamentarian:* Mr. Speaker, I can only think that the honourable member opposite will die of the pox if not on the gallows!
> *Second Parliamentarian:* I should imagine that the dreadful contingencies to which my friend alludes depend entirely upon whether I embrace his mistress or his principles.
> *Mr. Speaker:* Gentlemen, gentlemen! Order, please.
> *First Parliamentarian:* Abject apologies, sir, I have offended the House.
> *Second Parliamentarian:* And I, sir, have offended truth; I am repentent!

The fourth temptation of the debate is ambiguity. We have drawn our example from Aristotle's early work on logic, *Sophistical Refutations*. It is in the form of a very simple argument.

> There must be sight of what one sees.
> One sees the pillar.
> Therefore, the pillar has sight.

One can easily see that subtler ambiguities can pervade many a debate.

2. The Debate: Its Evaluation

The logical evaluation of a debate is somewhat more complicated than the evaluation of the quarrel. The question is explored more fully in

Chapter Six, when we say something about the fallacies. This much now can be said:

1. The principal object of a debate is to win a majority vote or judgement.
2. This fundamental object may or may not be compatible with getting at the *truth* of the matter at hand.
3. When a debate serves the course of proof and reason, it can be one of truth's and reason's most effective allies.
4. But it must be emphasized that the primary object of debate is to *win*, even if truth and reason must be sacrificed or set aside temporarily.

In its barest essentials, the debate is no friend of reason; logic can only advise that debating techniques, as instruments of reasoning, be used with great caution.

One of the major theoreticians of argument in the nineteenth century – and one of the modern world's seminal political thinkers – was John Stuart Mill. Mill's classic little book, *On Liberty*, contains his worry about debates, especially as tools of political decision-making.

Debates occur in Parliaments, or Congresses, Diets, Knessets, Bunds, or other such forums. (From now on we shall use the terms "Parliament" and "Parliamentary" generically, to refer to any such forum or any such procedures or practices.) The members of Parliament are individuals; each has his or her own intelligence, experience, opinions and prejudices. Debates make contending and contradictory claims on these individuals for loyalty and support.

In any debate, the contending debaters, who represent all sides of the question, put the significant proposition, and all alternative propositions, to the test. Ideally any member of the House can rise and ask probing questions or offer his or her own advice or criticisms. As the debate unfolds, certain alternatives drop from contention; no one will now support these alternatives, for they have not passed the test of inspection or scrutiny by all sides of the House. Sooner or later, a consensus (or something close to it) is reached. The consensus will probably support none of the original propositions in contention; instead, it will support some skillfully made compromise, tempered by the fires of debate. Usually what is endorsed is the proposition most likely to be true and most likely to be worthy of political support and action, for it alone has survived the furnace of contention and the onslaught of dogmatism and personal prejudice.

What an interesting and ingenious defence of the debate as a logically sound instrument of responsible reasoning! Its most original and intriguing aspect lies in this assertion:

> Admittedly, the objective of any given individual debater may well be to get his or her own way, even at the expense of objectivity and truth. In a community, there may be many such individuals, each having his or her private objectives uppermost in mind. Provided there are procedural rules, which enable the

> contending issues to be freely and openly examined by all concerned, private prejudice will more often than not defer to the general will. Though an individual debate is no basic friend of truth, as a community or group activity the debate is an effective and objective *route* to truth.

Mill's optimism is refreshing and no doubt occasionally well justified. It is an optimism that is reflected in two of the modern world's most prominent ideas. One is the free market model of a commodity's value. Imagine an economy in which there is a market of consumers, access to which is open to all who would wish to sell. The consumers furnish whatever degree of demand there may be for any given item offered for sale, and the supplier and the sellers determine the supply. In the competition of various goods for the limited resources of the consumer, the laws of supply and demand ultimately determine how the contest is to be won, and what value is to be accorded each commodity. The worth of a commodity is determined by the degree to which it is well-received or approved by the consumerate.

The other idea that we can see reflected in Mill's defence of Parliamentary debates is the nineteenth-century evolutionary concept of the survival of the fittest. In evolutionary theory, a species is routinely subjected to stresses and afflictions, which test its capacity for biological continuation. It is a biological commonplace that a species is eliminated when it cannot adjust to such stresses, and that those species that do adjust survive. These are judged by Nature to be the fittest.

What Mill is offering us, then, is a kind of sematic free-enterprise and survival-of-the-fittest model – and justification – of the debate. We use the word "semantic" to indicate that responsible reasoning is concerned with the discovery and pursuit of the *truth*. In a logician's parlance, it is customary to say that concern with truth is a *semantic* concern, and we shall adopt that terminology here. Thus, the *semantic* free-enterprise and survival-of-the-fittest model of debate is one in which *truth* is most valued in the free marketplace of ideas; truth best survives the destructive forces of opposition and criticism. Mill's point is markedly similar to one made by Aristotle, centuries earlier. Aristotle was not overly concerned with the subjective emphasis of mere advocacy, since, as he thought, "truth and justice have a *natural tendency to prevail* [Italics added]."

3. Mill's Model and the Courts: A Sub-Model of the Debate

Perhaps the most impressive contemporary example of Mill's conception of free competitive argument as a road to truth is our judicial system. The best examples are the Criminal Courts and, more particularly, the institution of trial by jury. The basic format of the trial by jury closely resembles that of the Parliamentary debate. The judge presides, interprets procedure and rules on matters of Law, and so is a counterpart of the Speaker of the House. The prosecution and defence counsel are counterparts of the debate's contending parties. The jury is the counterpart of the House itself, or a debate's panel of judges. There are crucial differences, however. For example, the trial introduces two new elements into the structure of argument. In the trial, we have a new kind of participant, *the witness*, who gives evidence under oath; and his oath commits him to tell the objective truth (not just to state his party's, or his *side's*, position). Sworn testimony is, in the Courts, the sole access to matters of fact.

We can see that the problem of premissory instability takes on a new twist. In the quarrel, premissory instability is a serious problem and it makes for argumentative anarchy; in a Parliamentary debate, premissory instability among the contending participants is no problem whatever. The Courts, however, are charged not merely with the responsibility of reaching a decision by the vote of a jury; that decision must be based upon as scrupulous a presentation of *the objective facts* as circumstances allow. It is, then, a distinctive feature of the criminal trial that it has devised procedures for the accumulation of *factual evidence* that reasonable men and women could tend to agree upon. The Courts respond to this requirement ingeniously, by reserving the evidence-producing aspects of a trial for those who are *not* contending participants in any trial. Thus neither the lawyer nor the judge is the arbiter of the facts; only the jury is. In a trial, the obligations of as much factual objectivity as the circumstances allow are not placed on the shoulders of those who are actually doing the arguing.

There is a second new element in the argumentative structure of the trial. Correlative with the notion of evidence is the *requirement of proof*. Any contention before a Court is settled only by proof. A trial tries and proves a proposition, and therefore demonstrates the truth or falsity of that proposition. The overriding idea is that, in an orderly, open and adversarial competition, freely and impartially judged and fairly presided over, the truth will out. Again, we see an expression of semantic free enterprise and survival of the fittest.

Recall Aristotle's confidence that truth and justice have a natural tendency to prevail. But compare this with the pessimism of René Harding in Wyndam Lewis's angry novel, *Self Condemned*.

> For honestly you must grant me this, that no individual could be guilty of the follies that most bodies thousands strong, which we call "governments", are guilty of. You will object that what the governments have to handle are far more complex matters than what a man would have to cope with. But "complexity" is no excuse really for stupidity. Most of the things statesmen have to deal with are fundamentally as simple as the running of a hotel. . . . The fact

remains, however much one may argue, that only one man in a hundred thousand turns out to be a murderer – and he ends his life on a gallows or the "hot squat". Whereas there is no civilized nation that finds itself a proper nation until it has taken human life, to the tune of a million or so. If you murder enough people it's all right.

4. Critique of Mill's Semantic Free-Enterprise and Survival-of-the-Fittest Model of Debate

It may be that abstractly and ideally considered, Mill's model might be as good a bet as to the reliability of an argument as one could think of. Practically speaking, however, it can be seen to be a rather forlorn hope.

There are some rather sobering facts to recognize about how human beings act and fail to act in supposedly free and open group decision-making or judgement-forming situations. Consider just one example, the so-called bandwagon effect.

In recent years, the social technology of group decision-making has confirmed something that many people had already tended to believe on impressionistic and unscientific grounds: a great many men and women are prepared to go to extreme lengths to avoid being losers. If it appears that the momentum in a committee is likely to carry my position to the losing side of the vote, then I may switch my allegiance in order to avoid defeat. This, in turn, could serve as a model for others, who may be wavering and who may also hate to lose, especially if I am a person of any standing or reputation on the committee. And before you know it, everybody has "jumped on the bandwagon", ignoring the actual merits of the case at hand.

Other factors impinge as well: the desire to please one's colleagues; the disinclination to prolong the meeting unduly by holding out too long for one's own point of view; the unwillingness to bear up indefinitely under an unyielding and vocal opposition; the worry that one is just being pig-headed. No one of these is anything to be ashamed of, but all these feelings are vulnerable to exploitation; they are all there to be appealed to and hidden behind, even when the *real* motivation for switching (changing one's mind) is that one can't stand being on the losing side.

Another difficulty is that Parliaments, Congresses and Legislatures are not, by any stretch of the imagination, free markets. Rarely, in its contemporary operation, is a Parliamentary debate open to all members. Usually a debate has a time limit, and it cannot be guaranteed that the issue at hand will receive sufficient attention to produce a fair and accurate verdict. Many sessions of Parliament – most, in fact – are not well attended. This is not an indication of the laziness of parliamentarians; it shows their recognition that most debates are *faits accomplis*, that is, foregone conclusions. Why should this be so? It is mainly because the give and take of a debate is not free. Members are bound by party and caucus discipline; they are not free to be rationally persuaded into a position; and they are beholden to the Party Whip, that is, to the member of the party responsible for seeing that members adhere to the party line.

The courts do not exemplify the same deficiencies. But here, too, there are problems. Though eye-witness testimony is usually supposed to give the best possible access to the facts of a case, such perceptions are often unreliable. Any number of investigations into the psychology of eye-witness perception disclose that eye witnesses are often bad witnesses.[1] One problem is the witness's reaction: if the event witnessed at first-hand was of a shocking criminal nature, the shock and fear of the witness will interfere with the efficiency of his perceptual data-processing mechanisms. Even if his memory is in good order he may, without knowing it, be remembering what he originally misperceived.

There is another difficulty. Often the essential evidence in a trial deals with the barest and dullest of details, which the ordinary observer would not normally see or remember. But surely, it may be suggested, if the witness saw what happened then he *must* know whether the accused was wearing a bow tie. Experiments indicate that eye witnesses are extremely sensitive to the idea that, if they saw something happen, they will be acquainted with all or most of the perceptible details of what went on. In fact, no theory of perception supports such a notion, and there is some reason to think that eye witnesses tend to be *arbitrarily suggestible* about small details. That is, they invent details that they did not in fact perceive, unwittingly and under pressure of the unwarranted assumption that eye witnesses record, process and can retrieve virtually every perceptible aspect of what they have observed.

There is a third difficulty. Since direct eye-witness testimony is not always the best route to a knowledge of what really occurred, courts often place their confidence in the indirect, inferred testimony of experts. But reliance on the authority of expertise can be risky. First: expert witnesses are identified as witnesses *for the prosecution* or *for the defence*. This tends to encourage the idea that expert witnesses are, at least to a degree, *partis pris*, that is, prejudiced in favour of one side or the other. Second: arguments that depend on an expert's so-called authority expose themselves to the possibility of the fallacy of *ad verecundiam*. The *ad verecundiam* basically involves a deficient appeal to or use of the expertise of an authority in support of a given conclusion. In very broad strokes, the fallacious "appeal to authority" may take two forms. First: the appeal might cite an authority whose quite genuine expertise offers no legitimate support of the conclusion at hand. For example, Bobby Hull is one of the world's finest hockey players. His television commercials for Philco products involve an *ad verecundiam:*

we are meant to believe that his hockey expertise qualifies him to speak authoritatively for electronics. Second: the appeal to authority can also involve reference to an expertise that simply does not exist. For example, Sue is absolutely moonstruck over good old Bill. She just might find herself saying, and believing, "Look, it's as clear as clear can be. The nation-state is outmoded, and will be replaced by the year 2050 with a world federalist structure; my boyfriend, Bill, told me so!" We shall say more about the *ad verecundiam* in a later chapter.

In summary, the criminal trial is subject to three major problems:

1. Group decision making can be notoriously unstable and non-objective
2. Eye-witness perceptions are often incomplete and unreliable
3. Expert testimony can be non-objective and can involve the fallacy of the *ad verecundiam*.

5. The Ad Populum

We noticed that in the debate and in the criminal trial, argument is directed to a specific group of persons – the referees, juries, judges or other deciding parties. The argument is not exclusively an attempt to argue from true premises. In such person-directed argument, there is a systematic danger of the *ad populum* fallacy, the fallacy of appealing specifically to the sentiments or prejudices of the group that the argument is designed to persuade. In arguing *ad populum*, when selecting premises it matters little whether they are true. The question is rather one of whether the premises are plausible to the audience and will be accepted, if possible enthusiastically, by that specific audience.

In this connection there seems a clear parallel between the *ad populum* and the *ad hominem*. Of course there is a difference of orientation in that the *ad hominem* is negative in its intent to discredit, whereas the *ad populum* is positive in its intent to win approval. But the subjective element of the selected premises is common to both.

More needs to be said about evaluating popular appeals, but familiar examples of excessive appeals to popularity are not hard to find. Consider the following dialogue:

> *First Candidate:* If your position is that physicians should not be allowed to opt out of medicare, and must be bound in their practices entirely by government-set standards and rates, how can you defend the declining standard of health care that is resulting from the current emigration of the best qualified physicians from the province?
>
> *Second Candidate:* I was born and raised right here in

Winnipeg, and what's clear to me is that the people of
this fine province have a right to medical care when
family members are in desperate need. I know that the
people support this fundamental right, regardless of
the travel whims of a lot of fancy, overpaid specialists.

At least one thing that appears to have gone wrong with this reply is
that it fails to answer the question of whether the emigration of
physicians will, in fact, cause a decline in standards of health care in the
province. The strategy is to try to reply effectively by a popular appeal in
place of examining evidence relevant to the point at issue.

Many familiar illustrations of the *ad populum* are provided by
commercial advertisers who, instead of arguing that their product does
what it is supposed to, try to convince a target audience that
trendsetters or other popular models of their group use it, so that anyone
who does not use it is out of the mainstream. A well-known hamburger
chain shows lovable children, parents and grandparents indulging in
joyous scenes of family solidarity and other lovable but ordinary daily
activities. The staff making the hamburgers are youthful, efficient,
friendly girl- or boy-next-door models. By the end of the commercial, one
feels that it would be an act of almost patriotic family solidarity to eat at
one of these restaurants. But has this message told us anything about
the nutritive value of their product, or other facts that might play a rôle
in rationally deciding whether to dine there? It is easy to suspect that
much of the appeal is simply *ad populum*.

Exercise

Show whether the example of insincerity from *Julius Caesar*
exhibits a fallacy at any point.

But what is fallacious about the *ad populum*? Is it not sometimes
necessary, if you are to win your case, to select premises that your
audience is likely to accept? In debate, it would be pointless to do
otherwise. What could be fallacious about appealing to beliefs of your
specific audience? The answer is that there is always grave danger of
fallacy in the outright partisan process of trying to convince a target
audience by utilizing whatever assumptions, no matter how outrageous,
that audience shows its preparation to tolerate. Such a subjectively
oriented strategy can subvert the objective goal of arriving at the truth
by means of logical reasoning. But what is the specific fallacy?

We can see that there are two fallacies involved. The first fallacy
implicit in the *ad populum* is the same error as that of the *ad hominem*,
namely making the inference to the truth or falsity of a proposition
based on the subjective circumstances or beliefs of a particular person or
group. Of course, hardly anyone is fooled by this form of reasoning in its

most clear and simple form, as expressed in the pair of arguments below.

Everyone believes it.
Therefore, it is true.

Nobody believes it.
Therefore, it is false.

These are transparently incorrect forms of argument. But in the turbulent emotions of political solidarity, of mass enthusiasms, of the struggle for ascending of interest groups and lobbies, such specious arguments can come dressed in protective clothing.

Exercise

Evaluate the following pair of arguments. If there is a fallacy, show how it is committed.
(1) We're all working men here. I know what it is to put in a day's work like the rest of you. That's why we must resist any attempts to legislate fixed-wage guidelines. The cost of food and clothing keeps going up. Consequently, fixing wages is a policy that will strangle the average family.
(2) We're all businessmen here, and we know the value of a dollar. That's why we must have some form of wage guidelines before the private sector is strangled by unrealistic wage escalation and the economy comes crashing down. Not fixing wages is a policy that can ruin the financial prospects of every family in the nation.

6. A Second Form of Ad Populum

We saw how *ad populum* can be an incorrect type of argument. But like its partner the *ad baculum*, the *ad populum* is often an attempt to waive, bypass or obscure argument altogether. In such instances of the *ad populum*, it is difficult or perhaps even impossible to find propositions making up a set of premises at all.

A well-known lumber company makes successful use of the popularity of the "back-to-the-land" and "do-things-for-yourself" North American folklore, especially popular in the "counterculture" of recent years, to appeal to a wide stratum in its population of possible buyers. Rather than speak of the quality of the tools, lumber, or other products it advertises, its commercial message conveys feelings of accomplishment and pride that are associated with building things for yourself. The point has been evaded, true, but the advertisement hits a popular target by appealing to the pride of personal do-it-yourself accomplishment. Perhaps, then, something evasive or manipulative has transpired. But where is the fallacy? Indeed, where is the argument? A large part of what seems to be wrong is that argument of any sort has been forgone in favour of a direct and quite successful appeal to widespread sympathies and attitudes. Where are the premises and conclusions in such an appeal?

The fallacy here is not to be found in the incorrect use of argument, but rather in the manoeuvre to short-circuit rational argument by jamming it with emotional interference. Why trouble to mount logical arguments if you can arouse your target audience so much more directly and winningly by appealing directly to raw emotions? In the lumber company example, the mass emotional impact obscures our attempts even to identify clearly what the argument is supposed to be. It is as if the advertiser were saying, "Forget the dull business of evaluating the evidence concerning the reliability or efficiency of tools. Love them for the noble ideals they represent. This is so much more fun anyway." Thus the *ad populum* is effectively used by advertisers who sketch rich fantasies of mass approval and admiration. A fantasy may make a "statement" of sorts, but can we assume that it is an identifiable set of propositions that can be evaluated as true or false? If not, the fallacy is more one of distraction than incorrectness of argument.

Practically speaking, hard argument is often mixed in with soft, non-argumentative appeals. As Jacques Ellul points out, the overall effect of propaganda can be heightened by mixing genuine information with pure propaganda.[2] In advertisements for automobiles or electrical applicances there are often legitimate facts about technical specifications or proved performances added to appeals to feelings and passions. It is well known also that the Nazi master of propaganda, Josef Goebbels, often included much genuinely truthful information in otherwise exaggerated and heavily emotional newsreel reports from the battlefront. He realized that the factual elements greatly escalated the credibility and value of the propaganda, which might otherwise be hard to swallow. Consequently, in dealing with the *ad populum*, the first step is to clearly identify the premises of the argument and try to sort them from their emotional surroundings. Sometimes, as we have suggested, there aren't even any clearly locatable premises in an *ad populum* appeal.

Exercise

Search through newspapers, magazines, or other sources for some suitable examples of "hot" rhetoric or debate. Mark what you think are the premises and conclusions of the argument, giving reasons for your interpretation. Then indicate any parts of the text that you can identify as exhibiting some specific type of *ad baculum, ad hominem*, or *ad populum*.

Summary

The quarrel and the debate are of some interest as models of argument that can illustrate the cut and thrust of objection and reply. The debate especially can be a valuable source of argument; it reveals hidden presuppositions or throws new light on a subject by argument. But we have seen a number of practical shortcomings of both the debate and the quarrel as models of argument that can be conducive to truth and reason. Among the most impressive shortcomings of the quarrel is its almost inevitable tendency to descend to the level of the *ad hominem* and *ad baculum* fallacies. As for the debate, it is too often open to the *ad verecundiam* fallacy. Even worse, its audience-directed, adversary structure causes it to be, by its very nature, continually open to the *ad populum* fallacy.

As a consequence of these built-in practical limitations of the quarrel and the debate, we shall move on to other models of argument that are less subjective in nature.

Notes

1   For a disturbing case against the reliability of eyewitness testimony, see E.F. Loftus, *Eyewitness Testimony* (Cambridge: Harvard University Press, 1980).
2   Jacques Ellul, *Propaganda* (New York: Knopf, 1972).

Chapter Three
The Deductive Method

The problem with the quarrel and the debate is that they tend to let concern for truth and cool reason be submerged; they are concerned only with winning. Seeking a more solid foundation for argument, we turn to the scientific approach, with its objective method. Mathematical logic is a part of science that would seem to be a likely source of firm and objective guidelines to evaluate arguments generally. In this chapter, we will outline the basics of standard – or so-called "classical" – elementary deductive logic. Along the way, we will try to assess the values and limitations of this classical deductive logic as a tool or component in evaluating arguments.

1. Entailment

In the quarrel and the debate, the arguments we have been considering have been *activities* between human beings, activities directed to the dynamic processes of reasoning. But there is a different way to look at an argument. In the deductive model of argument, an argument is a set of propositions, one of which is already designated as the conclusion, and the others as the premises. This set of propositions is fixed in advance, already given. The only task we need concentrate on is to decide whether the relationship between premises and conclusion conforms to standards of deductive correctness.

Taken in this way, an argument need not involve any exchange between persons; it could be merely written down on a blackboard or on the back of an envelope or printed in a book. From now on we will feel free to use the word "argument" in either of these senses, depending on context. While it is perfectly true that arguments of the "impersonal" sort could be actual, personal exchanges, it is also true that they can be composed, and can be studied from outside the context of give-and-take. In the *purely deductive argument*, we are exclusively preoccupied with the presence or absence of *one specific relation* between the premises, on the one hand, and the conclusion, on the other. This relation logicians call *entailment*.

Entailment can be defined simply as follows: *a proposition,* p, *entails a proposition,* q, *if and only if it is not possible that* p *is true and yet that* q *is false.*

Thus *p* will entail *q* when, given that *p* is true, *q* also must, *of necessity*, be true. In such a case, we say that there is a *deduction of* q *from* p or that q *is deducible from* p. Example: "three is greater than two" (*p*) entails "two is less than three" (*q*). The first phrase, *p*, entails the second phrase, *q*, because it is not possible that "three is greater than two" could be true while "two is less than three" is false.

It is easy to see that purely deductive arguments are often objects of study, in such disciplines as mathematics and mathematical physics, where the notion of deduction is clearly tied up with the notion of *exact* or *strict* proof.

If you take a purely deductive interest in an argument, then you need not concern yourself with whether the premises are true, or whether the conclusion is true. Your emphasis is entirely on the question of the *relationship* between premises and conclusion: *if* the premises are true, is it or is it not the case, then, that the conclusion must, as a matter of strict necessity, also be true? *That* is the question you ask if you are taking the purely deductive point of view. Such arguments, therefore, can be perfectly good, pure deductions of their conclusions even when their premises are false or silly or preposterous.

Obviously, arguments involving purely deductive considerations are of rather selective or restrictive interest. In particular, an argument can be an excellent pure *deduction*, but nevertheless a bad *argument*. Here is an example.

    (1)   If all bachelors are rich, then all cars are fast.
    (2)   All bachelors are rich.
    (3)   Therefore, all cars are fast.

In this argument, as in all deductive arguments, deductive correctness (entailment) is a conditional concept. This means that *if* the argument is correct, and *if* the premises are true, the conclusion must also be true. But it is not necessary for the premises all to be true (or even for any of them to be true) for the deduction to be correct. It is merely necessary for the conclusion to be true *if* the premises are true. The actual truth or falsity of the premises is not a matter of interest from a logical point of view. As mathematical logicians, we are interested only in the relation between the premises and the conclusion. We are interested only in whether the premises entail the conclusion, in other words, in whether the deduction is correct. And it is not necessary that the premises be true in order to have a correct deduction. In the example above, suppose that both premises are false. Does this affect the correctness of the deduction? No. Why? Because deductive correctness is a conditional concept. Thus, to say that a deduction is correct is not to say that the argument is, in all respects, acceptable. It is only to say that the conclusion is true *if* the premises are true.

To say that *p* entails *q* is to say that it is logically impossible that *p* be true and *q* false. If we say that "two is greater than three" implies "three is less than two", we mean that it is *logically* impossible that two be greater than three while three is not less than two. To say that two is both greater than three and three is not less than two is *logically self-contradictory*, or *inconsistent*.

Another way to characterize *logical inconsistency* is to contrast it with two other similar but distinct concepts, *physical impossibility* and *physical improbability*. We will take a man, George, as an example. We might say of George that it is improbable or unlikely that he would run a mile in six minutes. If we found out that George was a hundred years old and had arthritis, we might even say that it is physically impossible for George to run a mile in six minutes. But notice that, in neither of these cases, is it *logically* impossible for George to run a mile in six minutes. It is not self-contradictory that George should run a mile in six minutes. It is possible to imagine a one-hundred-year-old, arthritic George, running a speedy mile in six minutes. The proposition, "George, who is a hundred and has arthritis, just ran a six-minute mile" is not a logical impossibility, and does no violence to the logical necessities.

But here is a logically impossible proposition: "George is both taller than and shorter than Albert." Taken literally, and not as a figurative expression or a joke, this utterance expresses a logically impossible proposition. To say that George is simultaneously taller than and shorter than Albert is flatly *inconsistent*.

In general, then, we must distinguish between logical impossibility and the kindred concepts, physical impossibility and physical improbability.

Exercises

1. Which of the following assertions involve logical impossibilities? Which involve physical impossibilities? Which involve improbabilities?
   a) Henry is seven years younger than his own younger brother.
   b) Henry is sixty years older than his own younger brother.
   c) Henry is the same age as his six brothers.
   d) Zelda is married to a bachelor.
   e) Zelda is married to a twelve-year-old boy.
   f) Jack drew a square circle.

2. An explorer has reached a river, which is one of the boundaries of an exotic country that he wishes to make a study of. In this country there are only red-haired persons and brown-haired persons. It is impossible for people with red hair to tell anything but the truth; it is impossible for people with brown hair to tell anything but lies. It is now sunset, and our explorer sees, across the river, the figures of three men, members of this strange tribe, silhouetted against the sinking sun. Understandably, he cannot make out the colour of their hair. The explorer and the men have the following conversation.

Explorer: Hello there, will you please tell me which you are, red-haired or brown-haired?

First Native: (Because the First Native spoke so softly, the explorer did not hear his reply.)

Second Native: He says he is red-haired and he is red-haired.

Third Native: Don't be silly, he is brown-haired.

This conversation gives all that is necessary to compose a pure deduction that correctly identifies the hair colour of each of the three natives. The deduction is sketched below. See if you can compose the deduction without looking at the answer.

Purely Deductive Solution to the Native-Identification Riddle.

    The explorer asked, "Are you red-haired or brown-haired?" If the first speaker were red-haired then, being a truth teller, he would, of course, say that he is red-haired. If he were brown-haired, he would, of course, lie and say that he is red-haired. So you know that there can only be one answer to the question put to the first speaker, namely that he is red-haired.

    The second speaker asserted that this was, in fact, the first speaker's reply. So the second speaker told the truth, and therefore is red-haired. Since he is a truth-teller, he also tells the truth when he asserts that the first speaker is red-haired. Consequently, we know that the first two speakers are red-haired.

    But the third speaker claims, falsely, that the first speaker is brown-haired. The third speaker is a liar, and is himself brown-haired.

2. Conjunction, Disjunction and Negation

The structure of classical logic is determined by the *connectives:* ∧

(and), ∨ (or), ⅂ (not). These connectives join propositions together to make larger propositions. The connectives are called *constants* because their meanings are firmly and exactly fixed by the definitions below. Letters used to stand for propositions (*p*, *q*, *r*, . . . and so on) are called *variables* because a letter can stand in for any proposition. The only requirement is that, once *p* has been used to represent some proposition in an argument, the next time that same proposition occurs, it must again be represented by *p*. "*p* ∧ *q*" *(p and q)* is said to be true where *p* and *q* are both true. If either one is false, the whole proposition "*p* ∧ *q*" *is false.* "*p* ∨ *q*" *(p or q)* is said to be true where at least one of *p* or *q* is true. It is only false if both *p* and *q* are false. "⅂ *p*" (not-*p*) is true where *p* is false, and false where *p* is true. That is, "⅂ *p*" always has the opposite truth value of *p*. It is called the *negation* of *p*.

This information can be summed up in the following truth-tables.

Conjunction (and)				Disjunction (or)				Negation (not)	
p	*q*	*p* ∧ *q*		*p*	*q*	*p* ∨ *q*		*p*	⅂*p*
T	T	T		T	T	T		T	F
T	F	F		T	F	T		F	T
F	T	F		F	T	T			
F	F	F		F	F	F			

As we will see, sometimes "*p* or *q*" means "either *p* or *q*, but not both." Such cases will be defined differently from ∨, above. A concrete illustration may be helpful. Circuit A, where the switches are in a row, behaves conjunctively; circuit B, where the switches are parallel, behaves disjunctively.

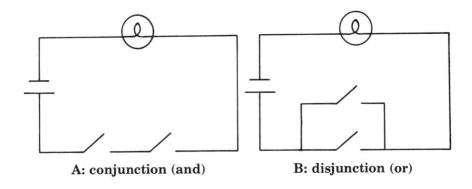

A: conjunction (and)   **B: disjunction (or)**

In circuit A, the light goes on only if both switches are closed. If either switch is open, the light will be off; if both switches are open, the light will be off. In circuit B, the light will be off only when both switches are open. If either switch is closed or if both switches are closed, the light will be on. In A, both switches must be closed to turn the light on, whereas in B, only one switch need be closed.

Each of the symbols ∧, ∨, and ⌐ is a *truth-functional* connective. That is, any compound proposition, no matter how large, can always be shown to have a truth-value that is fixed once the truth values of all the component-propositions, p, q, r, and so forth, are fixed. In mathematics, a function is something that yields a definite value if definite values are put into it. For example, n^2 is a function because, for any number you put in for n (say, 3), you will get a definite number determined by that function (in this instance, 9). That is the beauty and simplicity of classical deductive truth-functional logic. Every possible outcome is determined simply by the truth-table definitions of the connectives.

To see how truth-functionality works, we need to see how parentheses can be used to provide an exact and unambiguous procedure for making bigger propositions out of small ones. For example, suppose that we want to form a disjunction made up with a conjunction, as in this expression: p and q or r. This phrase could have two meanings: (p and q) or r; p and (q or r). This is a problem because we can see that the two expressions have different truth-values, which depend on the truth-values of the components. Let us start with $(p \land q) \lor r$. There are three variables involved; thus, there will be 2^3, or 8, different possible combinations of truth-values. The number of possible combinations can always be expressed "2^n" where n is the number of distinct variables. We can list all the possibilities in a table. We alternate pairs of truth-values under the variable at the right-hand side of the sequence of letters (under r in chart A). Then we list pairs, quadruples, and so on in succeeding columns as we proceed to the left, doubling Ts and and Fs each time.

p	q	r
T	T	T
T	T	F
T	F	T
T	F	F
F	T	T
F	T	F
F	F	T
F	F	F

A: Under r, we have alternated Ts and Fs. Under q, we have alternated pairs of Ts and Fs. Under p, we have alternated four Ts and four Fs. Check to see that the table represents all possible combinations of truth-values.

Next, we recall the truth-table definition for ∧. Thus the value for $p \land q$ must always be as in table B.

B:		*p*	*q*	*r*	*p* ∧ *q*	(*p* ∧ *q*) ∨ *r*
	(1)	T	T	T	T	T
	(2)	T	T	F	T	T
	(3)	T	F	T	F	T
	(4)	T	F	F	F	F
	(5)	F	T	T	F	T
	(6)	F	T	F	F	F
	(7)	F	F	T	F	T
	(8)	F	F	F	F	F

By a "row" we mean a horizontal line of truth values, for example, (1) in table B. By a "column" we mean a vertical line of truth values, like the sequence under *p* ∧ *q* in B above.

p ∧ *q* is true only in rows (1) and (2), because only in those rows are both component propositions true. Next we look to (*p* ∧ *q*) ∨ *r*. This statement is false only in rows (4), (6), and (8), because only in those rows is *p* ∧ *q* false and *r* also false. And we remember that the "or" is only false if both components are false.

By a similar procedure, we can construct the truth-table for *p* ∧ (*q* ∨ *r*), as in table C.

C:		*p*	*q*	*r*	*q* ∨ *r*	*p* ∧ (*q* ∨ *r*)
	(1)	T	T	T	T	T
	(2)	T	T	F	T	T
	(3)	T	F	T	T	T
	(4)	T	F	F	F	F
	(5)	F	T	T	T	F
	(6)	F	T	F	T	F
	(7)	F	F	T	T	F
	(8)	F	F	F	F	F

If we look at *q* ∨ *r*, we see that it is false only in the two rows where *q* and *r* are false. For that is the way the "or" is defined. And *p* ∧ (*q* ∨ *r*) is true only in the first three rows because only in those rows are *p* and *q* ∨ *r* both true. That is the way "and" has been defined.

By inspecting tables B and C, we find a difference of truth-value in rows (5) and (7). In those two rows, (*p* ∧ *q*) ∨ *r* *is true whereas p* ∧ (*q* ∨ *r*) is false. The parentheses make the difference in truth or falsity. If truth-functional logic is to be exact and unambiguous, we must use parentheses, or some similar device, to punctuate the sentences and remove ambiguities that could make a difference between truth and falsity, between correctness or incorrectness of an argument.

Exercise

Show whether there is a difference in truth-functional structure between ⅂(*p* ∧ *q*), which could be read as "not both *p* and *q*" and

⅂$p \wedge$ ⅂q, which could be read "not-p and not-q". Note that the negation sign applies to the proposition that directly follows it. In ⅂$(p \wedge q)$, the negation sign applies to the whole expression "$p \wedge q$". In ⅂$p \wedge$ ⅂q, negation applies only to its individual letter.

Some words that connect propositions do not have this property of truth-functionality. For example, the word "because" in "Bill died because he ate cod with chocolate sauce" is not a truth-functional connective: the truth-value of the compound sentence is not determined solely by the truth-values of the components "Bill died" and "Bill ate cod with chocolate sauce". Even if we know that both components ("Bill died" and "Bill ate cod with chocolate sauce") are true, we may still not know whether the compound is true, that is, whether his death was the result of eating cod with chocolate sauce. The connective "and", on the other hand, does seem to be a truth-functional connective. It seems to be truth-functional in the proposition "Bill ate cod with chocolate sauce and Bill died": the component values do seem to determine the value of the whole proposition.

Are the English expressions "and", "or" and "not" really truth-functional? The expression "it is not the case that" seems to be truth-functional. Applying this expression to any English sentence that is true or false results in a sentence with the opposite truth-value. The expression "it is not the case that" (or "not-" for short) and the connective ⅂ express the same truth-function.

It seems reasonable to expect that a sentence formed by joining *any* two true phrases by means of "and" will be true; if either or both of the original phrases is false, their conjunction will be false. For example, we have the two true phrases, "Bill punched Hank" and "Hank bled profusely from the nose." We might well think that "Bill punched Hank and Hank bled profusely from the nose" will also be true. Yet we might *not* be so sure about "Hank bled profusely from the nose and Bill punched Hank."[1] If we are not prepared to say that this sentence is true, then we cannot confidently claim that "and" and $\wedge$ express the same truth-function.

It is much the same with "or". In classical logic, the connective $\vee$ is such that the disjunction of any two false formulas is false, and otherwise is true. What of the counterpart expression in English? In general, it seems correct to say that whenever you have two sentences at least one of which is true, then their disjunction by means of "or" will also be true. But consider. Here are two true sentences (and so two sentences at least one of which is true): "Hank punched Bill" and "Gödel proved the incompleteness result." But not everyone will agree that "Hank punched Bill or Gödel proved the incompleteness result" is true, or, for that matter, even that "Hank punched Bill or Hank punched Bill" is true. Consequently there is some reason to think that "or" and the connective $\vee$ do not express the same truth-function.[2]

For present purposes, it is not necessary to decide one way or the other whether the English words "and" or "or" are truth-functional in some sense. It suits our purpose to raise the reasonable possibilities that they do not express the *same* truth-functions as their counterpart connectives in classical logic, and therefore cannot be said to have the same meaning as those counterparts. This ends the digression.

To sum up, a strong advantage of the truth-functional definitions of $\wedge$, $\vee$, and $\urcorner$ is always a precise way of calculating the truth-value of any proposition, no matter how complex, if the truth-values of its components have been determined. Let us take a complex proposition: $\urcorner (p \vee q) \wedge (\urcorner p \vee q)$. We know that p is true and q is false. We proceed systematically from the innermost components. Because p is true and q false, we know that $(p \vee q)$ must be true. Thus $\urcorner (p \vee q)$ is false. We know that p is true, so $\urcorner p$ is false. Hence $\urcorner p \vee q$ is false, because a disjunction is false if both components are false. Both parts of the proposition are false; hence the whole proposition must be false. Actually, we could have stopped after finding that $\urcorner (p \vee q)$ was false, because one false component is enough to ensure that the whole proposition, a conjunction, is false. A conjunction is only true if *both* components are true.

Exercise

Where p is true, q is false, and r is true, determine the truth-values of each of the following propositions:

(a) $(q \wedge \urcorner p) \vee (p \wedge q)$

(b) $(p \vee (q \vee p)) \vee (r \wedge \urcorner s)$

(c) $\urcorner\urcorner p \vee (q \wedge r)$

(d) $\urcorner(p \vee q) \wedge \urcorner(\urcorner q \vee \urcorner r)$

It is worthwhile noting that sometimes "or" is not *inclusive* like the $\vee$, which may be read as "either p or q or both". Rather, it may be *exclusive*, as described by the expression "either p or q but not both". If I say, "This number is either even or odd", my "or" is presumably meant to be exclusive. Whereas if I say, "A sore throat or sneezing is a symptom of a cold", we may presume I use the inclusive "or". Sometimes it is not clear which is meant, as in the statement, "He must be either drunk or insane". The truth-table for the exclusive "or" is like that for the inclusive "or", except that a statement where p and q are both true is excluded from being true. It could be defined as below:

p	q	p $\overset{\vee}{\vee}$ q (exclusive or)
T	T	F
T	F	T
F	T	T
F	F	F

Usually when logicians talk of disjunction, they are referring to the inclusive "or". $p \vee q$ could be defined as $(p \vee q) \wedge \daleth(p \wedge q)$, as you can see by constructing a truth-table for $(p \vee q) \wedge \daleth (p \wedge q)$. You will see that it has the same truth-value in all four cases as does the table for $\vee$.

3. Conditionals and Equivalences

The advantages of the truth-functional approach are simplicity and exactness. The approach provides a simple test for the correctness of any argument that can be expressed in the proper form. A possible disadvantage, which we have already noticed, is that, by limiting itself solely to the question of truth-values, it fails to take into account any connections between propositions other than their truth-value connections. As we saw, the sentence "Hank punched Bill or Gödel proved the incompleteness result" is true if either or both of its components are true. But because there is no apparent connection between the components, it seems to be stretching things a bit to say that the disjunction is true. Perhaps it is wrong to stretch things even a bit, however, if all we mean by saying "p or q" is that at least one of p or q is true – never mind whether they are connected in some way or not.

A *conditional* is expressed in the form "If p, then q" where the part after the if (the p, in this case) is called the *antecedent* (or assumption) and the other part (the q, in this case) is called the *consequent* (or what follows the assumption). Can we fit conditionals into the truth-functional mould? Well, the answer is that we can do it, but only if we stretch things more than just a bit.

Let's see how it might work. Consider this conditional: "If I throw this piece of chalk against the board, it will break."

	p	q	if p then q
(1)	T	T	T
(2)	T	F	F
(3)	F	T	?
(4)	F	F	?

The first two rows are fairly clear. If I throw the chalk and it shatters, then my conditional statement ("If I throw the chalk, it will shatter") may be said to be true. If I throw the chalk and it does not shatter then, as row (2) indicates, the conditional is false. But what if I don't throw the chalk at all, as in (3) and (4)? The best answer seems to be that the truth-value remains undetermined. But if we want a truth-functional conditional, the truth-value cannot remain undetermined if the values for p and q are fixed. What to do?

First, let's try making (3) and (4) both false under the conditional. The results are shown in column (B). But the results are the same truth-function as $p \wedge q$, and that can't be right! "*If* I throw the chalk, *then* it will break" is different from "I throw the chalk *and* it will break." What if we try column (C)? Column (C) gives us the same set of truth-values as q has, and that can't be right! "If I throw the chalk it will break" is different from "The chalk will break (whether I throw it or not)." What about column (D)? This would mean that "If q then p" would have the same set of truth-values as "If p then q." But that can't be right! "If I throw the chalk, it will break" is different from "If the chalk breaks I will throw it." So none of these possibilities can be tolerated.

We have now tried these three possibilities:

			(A)	(B)	(C)	(D)
	p	q	If p then q	If p then q	If p then q
(1)	T	T	T	T	T
(2)	T	F	F	F	F
(3)	F	T	F	T	F
(4)	F	F	F	F	T

What is left?

The only remaining alternative is to adopt the following definition. The truth-function so defined is called the *material conditional*, or "hook" ($\supset$).

	p	q	$p \supset q$
(1)	T	T	T
(2)	T	F	F
(3)	F	T	T
(4)	F	F	T

The relationship $p \supset q$ is false only where p is true and q false. This is the same truth-function as that expressed by $\urcorner (p \wedge \urcorner q)$. The idea that a conditional never goes from truth to falsehood is conveyed by row (2), the only row excluded from truth. But the new truth-function adds the idea that, if you start with a false proposition, your conditional will always be true.

What's wrong with this way of proceeding? First, it doesn't seem right, because even if the antecedent is false, the conditional could still be true. Suppose I do not throw the chalk: it might still be true that, if I throw it, it will break. Second, like the truth-functional "or", the material conditional can make the truth-value of the whole a function of the truth-value of just one part, even if the other part appears wildly unrelated. Thus the conditional implies that incommensurate propositions are true. For example, suppose it is false that John Woods has put his thumbs in his ears and repeated the word "gorilla" ten times. It follows logically that a conditional must be true: if John Woods has put his thumbs in his ears and repeated the word "gorilla" ten times, the theory of relativity is true. The only reasonable conclusion seems to be that "if-then" is simply not truth-functional. It could not be right that "if-then" means ⊃.

Research in philosophical logic has borne out the thesis that many conditionals are not truth-functional in nature. For example, according to some recent theories,[3] the conditional "If I were to throw this piece of chalk against the board it would break" is true *if and only if* there is a possible situation minimally different from the actual world – except that in this possible situation I *do* throw the chalk and it does break. The point is that a *hypothetical* situation in which I do throw the chalk must be different from the *actual* world in many ways, not in just this one single change. For in a situation in which I throw the chalk (unlike in the actual world, where I did not throw it), certain muscles in my arm would have had to be moved in a certain way. So the possible situation in question must be *as similar as possible* to the actual world, given this change in it. The logic of *the comparative similarity of possible situations* is not truth-functional, although it is based on classical truth-functional logic. Thus conditionals cannot be treated as exclusively truth-functional. To treat them as such ignores the whole business of possible alternatives and their logic.[4]

Other recent theories suggest that the truth-functional approach is too simplistic to serve as an exclusive approach to conditionals: it overlooks the whole question of how *p* and *q* are related to each other in "if *p* then *q*". These theories suggest that an apparatus must be added to truth-functional logic. The apparatus should acknowledge the different *kinds of relationships between propositions*, over and above their truth-values.[5] The device would acknowledge the connection between the subject-matter of *p* and the subject-matter of *q*, so that the connectives could not connect propositions that seem to be completely unrelated.

Each of these two areas of research suggests ways in which truth-functional logic can be enriched to enable us to deal more adequately with conditional arguments. But for our purposes, ⊃ will suffice. If we recognize the limitations of ⊃, and are clear that it really means nothing more than ⏋(p ∧ ⏋q), we will not get into trouble representing "if-then" by ⊃. Thus two limitations of truth-functional logic must be clearly kept in mind.

First, the material conditional does *not* tell us anything about the comparative similarity of non-actual – but possible – situations. (This is the job of another branch of logic, called modal logic.[6]) In particular, the ⊃ is not capable of handling the logic of counterfactuals, statements expressed in the form, "If *p* were true, then *q* would be* true," where the antecedent is always false. By the truth-table definition adopted above, the whole conditional in a material conditional is always true if the antecedent is false. But in a counterfactual – "If I were to throw this chalk it would break" – the conditional can be false where the antecedent is false. For example, suppose that, in the actual world, I do not throw the chalk. Even so, it might well be false that, if I were to throw it, it would break. Thus counterfactual conditionals are not truth-functional conditionals at all. Truth-functional logic does not handle conditionals with false antecedents as fully as we might like. For all that, we will see that deductive argument is a powerful tool of argument when it is harnessed to what is already known to be true. It is not applicable to all deductions in which we start with a false antecedent. But this does not alter its power. In its serious and careful use, it is an extremely effective method of reasoning.

Second, the material conditional does *not* tell us whether the subject-matters of propositions in an argument are connected. More about this limitation in Section Six.

The conditional is a one-way relation. "If p then q" is not the same conditional as "if q then p". "If you pay, you will eat" is not the same proposition as "if you eat, you will pay." But a conditional that does go both ways – called a *biconditional* – can be defined in terms of $\supset$ and $\wedge$ as follows: $p \equiv q = df (p \supset q) \wedge (q \supset p)$. The biconditional is translated as "if and only if". When we say, "Wood floats if and only if it is lighter than water," we mean that if wood floats it is lighter than water and if wood is lighter than water it floats. Let us construct a truth-table for $(p \supset q) \wedge (q \supset p)$.

p	q	$p \supset q$	$q \supset p$	$(p \supset q) \wedge (q \supset p)$
T	T	T	T	T
T	F	F	T	F
F	T	T	F	F
F	F	T	T	T

Using the definition of $\supset$, we construct a column for $p \supset q$. Then, using the same definition, we construct a column for $q \supset p$. Then we join the two columns with $\wedge$, and we use the definition of $\wedge$ to construct a column of truth-values for $(p \supset q) \wedge (q \supset p)$.

p	q	$p \equiv q$
T	T	T
T	F	F
F	T	F
F	F	T

We see that the biconditional is true if p and q have the same truth-values; it is otherwise false. Thus $p \equiv q$ means that p and q have the same truth-values. The statement $p \equiv q$ is defined by the truth-table immediately above.

Exercise

1. Where p is true, q is false, and r is true, determine the truth-values of each of the following propositions:
    (a) $p \supset (q \supset r)$
    (b) $(p \supset r) \supset (\daleth p \supset \daleth r)$
    (c) $(q \supset p) \supset (\daleth q \supset \daleth p)$
    (d) $[(p \wedge r) \supset q] \supset [p \supset (r \supset q)]$
    (e) $[(p \supset r) \supset p] \supset q$
    (f) $(s \supset p) \supset (r \supset q)$
    (g) $(p \supset q) \equiv (r \supset q)$
    (h) $(p \equiv r) \equiv (p \wedge \daleth q)$
    (i) $[(p \wedge q) \supset r] \equiv [p \supset (q \supset r)]$

2. Show that $(p \wedge q) \vee (\daleth p \wedge \daleth q)$ expresses the same truth-function as $p \equiv q$.

3. We have shown that $\supset$ is a one-way relation. It cannot be reversed or, as we say, it is not *commutative*. That is, $p \supset q$ is not the same truth-function as $q \supset p$. We saw, however, that $p \equiv q$ is commutative. Show whether $\wedge$ and $\vee$ are commutative.

4. See if you can find a truth-function containing only $\daleth$ and $\vee$ that is the same truth-function as $p \wedge q$.

4. Testing Deductions for Correctness

The truth-functional language can be used to test certain types of deductions for correctness. Remember that a correct deduction is one where it is logically impossible for the premises to be true and the conclusion false. To test this, we must first translate from English into truth-functional logic by replacing single propositions with letters, p, q, r, and so on, and replacing connectives, "and," "or," and so on, with their symbolic counterparts, such as $\wedge$. Once the deduction has been translated into standard symbolic form, a truth-table test for correctness can be applied. Consider this deduction:

(1) If 2 is a number, 2 has a successor
(2) 2 is a number

---

2 has a successor

Let *p* stand for "2 is a number"; let *q* stand for "2 has a successor". We translate the deduction into truth-functional language:

$$(1) \quad p \supset q$$
$$(2) \quad p$$
$$\overline{}$$
$$q$$

This deduction form is very common, and is called *modus ponens*. We can verify the correctness of the form by this truth-table:

	p	*q*	*p* ⊃ *q*
(1)	T	T	T
(2)	T	F	F
(3)	F	T	T
(4)	F	F	T

The deduction is correct if and only if it is logically impossible for the premisses to be true when the conclusion is false. We must look at the truth-table to see if there is a case where both premisses have the value T and the conclusion has the value F. In row (1), both premisses ("*p* ⊃ *q*" and "*p*") are true – and the conclusion, "*q*", is also true. In Rows (2), (3), and (4), at least one premiss is false. No case exists where the premisses are both true and the conclusion is false. We can therefore conclude that the deduction form, *modus ponens*, is correct; we also conclude that the original deduction, of which it is a translation, is also correct (assuming, of course, that the translation is accurate).

Another example:

(1) If Rocky throws the fight, Gloria will reject him.
(2) If Rocky doesn't throw the fight, the Cosa Nostra will take him for a ride.
(3) Either Rocky will throw the fight or he will not.

——————————————————————————

Either Gloria will reject Rocky or the Cosa Nostra will take him for a ride.

This deduction takes the form of a dilemma. Let *p* = "Rocky throws the fight", *q* = "Gloria will reject Rocky", and *r* = "the Cosa Nostra will take Rocky for a ride".

$$(1) \quad p \supset q$$
$$(2) \quad \neg p \supset r$$
$$(3) \quad p \vee \neg p$$
$$\overline{}$$
$$q \vee r$$

There are three distinct letters in this deduction form – *p*, *q*, and *r* – so there will be 2^3, or 8, possible combinations of truth-values these letters can be assigned. These truth-values are listed in the first three columns of the next chart.

	p	q	r	$\daleth p$	$p \supset q$	$\daleth p \supset r$	$p \lor \daleth p$	$q \lor r$
(1)	T	T	T	F	T	T	T	T
(2)	T	T	F	F	T	T	T	T
(3)	T	F	T	F	F	T	T	T
(4)	T	F	F	F	F	T	T	F
(5)	F	T	T	T	T	T	T	T
(6)	F	T	F	T	T	F	T	T
(7)	F	F	T	T	T	T	T	T
(8)	F	F	F	T	T	F	T	F

There will be 2^n rows in a truth table, where n equals the number of distinct letters in the deduction form. We set out all the combinations of truth-values for the single letters, p, q, r. We alternate Ts and Fs under r, pairs of Ts and Fs under q and quadruples under p. Then we have all possible combinations of truth-values represented.

Now we must check for correctness. Since there is no row in the truth-table where all the premisses are true and the conclusion is false, we conclude that the deduction is correct. In Rows (1), (2), (5), and (7) all the premisses – "$p \supset q$", "$\daleth p \supset r$", and "$p \lor \daleth p$" – are true; but in each of these rows, the conclusion is also true. There is no case where the premisses are all true and the conclusion is false.

One more example.

(1) If the Expos won the Cards lost.
(2) The Cards lost.

_____

    The Expos won.

Let p = "the Expos won"; let q = "the Cards lost". The result:

(1) $p \supset q$
(2) q

_____

  p

Unlike *modus ponens*, this deduction form is not valid. It is called the *fallacy of affirming the consequent*.

		p	q	$p \supset q$
	(1)	T	T	T
	(2)	T	F	F
X	(3)	F	T	T
	(4)	F	F	T

Note that there is a row of the truth-table where both premisses – "$p \supset q$" and "q" – are true and yet the conclusion, "p", is false. This row, marked with an X on the truth-table, demonstrates that the deductive form is not correct.

Exercise

Let p be "The Whips win their next game"; let q be "The Charlies win their next game"; let r be "the Mud Hens win their next game". Translate the following sentences into the truth-functional language.

a) If the Whips win their next game, then both the Charlies and the Mud Hens win their next game.

b) If the Whips win their next game, then both the Charlies and Mud Hens do not win their next game.

c) If the Whips win their next game, then the Charlies do not win their next game. But if the Charlies do not win their next game, then the Mud Hens win their next game.

d) If either the Whips or the Charlies win their next game, then the Mud Hens will not win their next game.

e) Neither the Whips nor the Mud Hens will win their next game unless the Charlies win their next game.

f) The Whips will win their next game if and only if both the Charlies and the Mud Hens do not win their next game.

g) The Whips will win their next game only if either the Charlies or the Mud Hens do not win their next game.

There is a third limitation of truth-functional logic; it is a limitation that applies to any formal logic. Once the deduction to be tested has been translated into the truth-functional language, checking for correctness is a purely mechanical procedure. However, the tricky part is translating from English (or whatever other natural language) into our "formal language". Since the English language is a many-splendoured thing, translation from it into an exact, formal language is sometimes difficult. Moreover, because English sentences are sometimes vague or ambiguous, and because we do not have an exact set of rules for English as we do for our truth-functional language, a sentence might be translated in several ways. Translation is more art than science. If the deductions being evaluated occur in a natural language, such as English, the mechanical procedures can only be applied *after* translation, and translation itself is not a mechanical procedure. But we cannot expect sentences of a natural language to be as exact as those of a formal language. Indeed, one of the most useful functions of logic, as developed formally, is to make arguments in English more precise. Formality has advantages; it also has limitations. Argument often occurs in natural language, even scientific argument. Hence we have found yet another significant limitation of classical deductive logic.

5. Summary: The Truth-Table Test for Deductive Correctness

Step 1 Set down the premisses of the deduction and number them. Then draw a line and set down the conclusion. Be careful to determine which proposition is meant to be the conclusion. Usually words like "therefore", "so", and so on indicate the conclusion. For example: "If the Expos or the Mets win their next game, the Cards will be out of the series. But the Expos win their next game if and only if the Mets don't win. So whatever happens the Cards will be out of the series." This argument becomes:

(1) If the Expos or the Mets win their next game, the Cards will be out of the series.
(2) The Expos win their next game if and only if the Mets don't win.

The Cards will be out of the series.

Step 2 Translate from English into the truth-functional language. Show which letter stands for which proposition. In this example, let p stand for "The Expos win their next game." Let q be "The Mets win their next game." Let r be "The Cards will be out of the series."

(1) $(p \lor q) \supset r$
(2) $p \equiv \lnot q$

---

r

Step 3 Make a truth-table by setting down columns representing all possible combinations of truth-values for all your propositions. There will be 2^n rows, where n is the number of letters in the argument. Thus if there are three letters, as in our example, there will be $2^3 = 8$ rows. Construct the rightmost column by alternating Ts and Fs, the next column to the left by alternating pairs of Ts and Fs, and so on, doubling the number of Ts and Fs each time. In our example, we will have three columns and eight rows.

	p	q	r
(1)	T	T	T
(2)	T	T	F
(3)	T	F	T
(4)	T	F	F
(5)	F	T	T
(6)	F	T	F
(7)	F	F	T
(8)	F	F	F

Step 4 Set down a column of truth-values under each premiss, using the definitions of the connectives. This will be easier if you set up a column for any components of premisses that are represented by more than one letter. In our example, we first make a column for $p \vee q$. The next column, which represents the premiss $(p \vee q) \supset r$, will be determined by using our semantics for $\supset$ between the $p \vee q$ column and the r column. In Row (1), for example, $p \vee q$ has the value T and r has the value T; thus $(p \vee q) \supset r$ will also have the value T (the semantics of $\supset$ being that a conditional is false only where the first element in it is true and the second false).

	p	q	r	$p \vee q$	$(p \vee q) \supset r$	$\daleth q$	$p \equiv \daleth q$
(1)	T	T	T	T	T	F	F
(2)	T	T	F	T	F	F	F
(3)	T	F	T	T	T	T	T
(4)	T	F	F	T	F	T	T
(5)	F	T	T	T	T	F	T
(6)	F	T	F	T	F	F	T
(7)	F	F	T	F	T	T	F
(8)	F	F	F	F	T	T	F

Having set down our truth-table, we can now test for correctness. If there is no case where the premisses are all true and the conclusion is false, then the deduction is correct. If there is any row – one or more – where all the premisses are true and the conclusion is false, then the deduction is incorrect. In our example, there is no row where the premisses are true yet the conclusion false; thus the deduction is correct.

Exercise

1. Translate the following arguments into the truth-functional language and show whether they are deductively correct.
   (a) Geometry is a part of either mathematics or physics. But it is not a part of mathematics. Therefore it is a part of physics.
   (b) If China invades Vietnam, the United States will start a nuclear war. But the United States will not start a nuclear war. Therefore China will not invade Vietnam.
   (c) Superman will die if and only if exposed to Kryptonite. It follows that, if Superman is exposed to Kryptonite, he will die.
   (d) If John moves his knight, Fred will move his queen. If Fred moves his queen, the game will be a stalemate. Therefore, if John moves his knight, the game will be a stalemate.
   (e) If John moves his knight, Fred will move his queen. So if the game is a stalemate, John moves his knight.

(f)   If John moves his pawn, his queen will be in jeopardy. If he moves his rook, he will lose it. But if he doesn't move his pawn, he must move his rook. Clearly, therefore, it cannot both be true that his queen will not be in jeopardy *and* that he won't lose a rook.

(g)   John will move his pawn only if Ernie moves his queen. But if Ernie moves his rook, John will knock over the board. It is not true that Ernie will move his queen *and* not move his rook. Therefore, if Ernie moves his queen, John will move his pawn *and* knock over the board.

6.   *Ignoratio Elenchi*

We saw that very often the *ad hominem* and the *ad baculum* fallacies use the tactic of avoiding the argument of one's opponent by introducing an irrelevant emotional or circumstantial appeal. Such a move is fallacious partly because it attempts to avoid or jam the argument by simply changing the subject. The fallacy is one of irrelevance, or the illicit changing of subject-matter. Aristotle called this kind of fallacy *ignoratio elenchi*, or misconception of refutation. He recognized that, when confronted by an argument one is unable to refute, a clever strategy is to mount another argument – any other argument that seems plausible, even if it doesn't respond to the first argument or even have anything to do with it. Often the audience won't notice the shift of territory, especially if the irrelevant counter-argument seems somehow related to the original one, or if it is colourful or emotionally interesting enough to make us forget about the first argument altogether. This fallacy is often called the *red herring*. In the sport of fox-hunting, a dried, smoked (hence red) and salted herring is drawn across the trail of the fox. This adds to the sport of the hunt by making things more difficult for the hounds.

Many interesting examples of this fallacy are given by R.H. Johnson and J.A. Blair.[7] They cite one illustrative case: (Senator) Paul Martin rose to defend his hometown of Windsor, Ontario against the remark of Arthur Hailey in his novel *Wheels*. Hailey had said that "grimy Windsor" is "matching in ugliness the worst of its U.S. senior partner [Detroit]." Martin is reported to have responded:

> When I read this I was incensed. . . . Those of us who live there know that [Windsor] is not a grimy city. It is a city that has one of the best flower parks in Canada. It is a city of fine schools, hard-working and tolerant people.

When he mentioned the flower parks, Martin did produce an argument to counter Hailey's appraisal. But then, as Johnson and Blair point out,

Martin changed the subject when he started to discuss the fine schools and hard-working, tolerant people. These facts may be appealing, but they are not relevant to Hailey's accusation; they do not tell us whether Windsor is ugly or fair. A shift of subject has taken place: the last part of the argument is an *ignoratio elenchi*.

We observed previously that the classical deductive logic of truth-functions cannot tell us whether subject-matters are connected. Thus classical deductive logic cannot tell us whether an *ignoratio elenchi* has occurred. We know that, in classical deductive logic, "p, therefore $p \lor q$" is a correct argument; so is "$\neg p$, therefore $p \supset q$" – even if p and q seem completely unrelated. This is a serious limitation of the application of deductive logic to argument. Classical deductive logic yields an account of "correct argument" that admits the possibility of committing the worst fallacies of irrelevance you can imagine.

We conclude that, while deductive logic takes us further, at least in one direction, than the debate or the quarrel, it, too, is open to fallacies; therefore, it is not a complete model of argument. The question remains, for later chapters, whether or how the deductive concept of argument can be extended to cope with its limitations.

Summary

Classical deductive truth-functional logic comes to grips with one core relationship between premises and conclusions: deductive entailment. A proposition, p, is said to entail another proposition, q, where it is logically impossible that p is true *and* that q is false. The classical logic of deductive entailment assures us that, if we start with premises that are true and known to be true, we will never arrive, by correct deduction, at a conclusion that is false. Thus classical deductive logic is a fail-safe procedure that never takes us from truth to falsehood; as well, it offers precise and simple methods for testing the correctness of any argument that can be expressed in its structure.

But classical truth-functional logic has significant limitations. (1) If we do not start with true premises – if we start with false premises, as in the counterfactual conditional – classical logic may not help us. Given the definition of $\supset$, $p \supset q$ is always true if p is false. (2) Many arguments occur in natural language, where fallacies can arise out of the ambiguities of words and sentences; classical deductive logic can not always deal with the idioms of natural language. (3) Classical logic does not take into account the possibility of changing the subject and committing *ignoratio elenchi*.

Within these limitations, deductive logic does provide a theory of some core elements of the logic of argument. In subsequent chapters, we will see if classical logic can be extended beyond the limitations outlined here to provide the essential structure of a fuller theory of argument adequate to the fallacies.

Notes

1 The interested reader might wish to pursue the discussion further. There is a paper by John Woods, "Is there a relation of intensional conjunction?", *Mind* LXXVI (1967), in which the answer is no; there is a critical reply by R.E. Gahringer, "Intensional conjunction," *Mind* LXXIX (1970).
2 An intersting discussion may be found in a paper by R. Jennings, "Or," *Analysis* 26 (1966).
3 See David Lewis, *Counterfactuals* (Oxford: Blackwell, 1973).
4 In a subsequent chapter, we explore this branch of logic, called *modal logic*.
5 See Richard L. Epstein, "Relatedness and Implication," *Philosophical Studies*, 36, 1979, 137–173.
6 See G.E. Hughes and M.J. Cresswell, *An Introduction to Modal Logic* (London: Methuen, 1968).
7 R.H. Johnson and J.A. Blair, *Logical Self-Defense* (Toronto: McGraw-Hill Ryerson, 1977), pages 55–64. Quote is reprinted by permission.

Chapter Four

Inductive Argument

Some of the trickiest fallacies crop up in inductive reasoning, especially where statistical claims or generalizations are made. It would be reasonable to expect that an inductive model of argument would play a rôle in helping us to classify and study these fallacies. However, as we will see, inductive logic is not as well developed as we might wish it to be; it is not a complete tool for the analysis of statistical fallacies.

One tool, however, *probability theory*, is useful. Despite its limitations as a model of argument, it can help us attain a somewhat clearer grasp of statistical arguments and their particular fallacies.

1. Induction: Its Description

The deductive argument and the inductive argument are both arguments in the second sense of the word "argument". That is, they are both groups of sentences; one of the sentences is the conclusion and the remainder of the sentences are the premisses. Our interest in inductive arguments will concentrate on the *logical relationship* between premisses and conclusion. What might such a relationship be, and how does it differ from the logical relationship of entailment, which we meet with in deductive arguments?

Remember that a *deduction* is a good one if its premisses *verify* its conclusion; remember that this mechanism of verification of a conclusion by its premisses provides *that the conclusion is true on condition that the premisses are true*. Finally, remember the relation that obtains between premisses and conclusions when the conclusion is true on condition that the premisses are true. That relation is called *entailment*.

For deduction, there are some fairly clear *interconnections*:

> Let p be a group of premises, and c a conclusion.
> Now consider the argument
>
> $$p$$
> $$\therefore c.$$
>
> This argument is a *good deduction* when
>   i) p verifies c,
> which means that
>   ii) c is true on condition that p is true
> which in turn means that
>   iii) p entails c.

Let us see whether there exists a similar pattern for induction. First, the basic idea behind an inductive argument is this: an induction is good to the extent that its premises increase the likelihood of its conclusion. Now let's look for a pattern:

> Let p be a group of premises and c a conclusion.
> Now consider the argument
>
> $$p$$
> $$\therefore c.$$
>
> This argument is a good induction if
>   i) p increases the likelihood of c

So far, we seem to have found a pattern that links induction and deduction. Both are arguments in which premises do certain things to the conclusion. But the premises do different things. Premises in deduction *verify*; in induction, they *increase likelihood*.

Condition (ii), in the deduction pattern, tells us what happens when premises verify conclusions: the premises make the conclusions true on condition that the premises themselves are true. Is there a similar second condition for induction? Yes. It is usual for logicians to say that the likelihood of a proposition is increased when its *probability* is increased.

In its essentials, the situation may be put this way. Every proposition is assigned a certain given, or initial, probability. The probability that it has will be represented by a number between 0 and 1. Thus there is a basic difference between a deductive argument – where every proposition, being either true or false, has an initial *truth-value* – and an inductive argument, where every proposition has an initial *probability-value*, or an *initial probability*.

If every individual proposition has its own truth-value, every *group* of propositions will have its own truth-value. Deductive logic tells you how to calculate the truth-values of any group from the truth-values of its member propositions. Similarly, inductive logic will tell you how to calculate *the probability of a group of propositions on the basis of the probability of the individual propositions that make up that group*. In an inductive argument, one can always calculate the probability of a group of premises if the probability of the individual premises is known.

In an inductive argument, the premises together will always have a certain *initial probability*, as the conclusion has its initial probability. An induction, "$p, \therefore c$", is a good one if the probability of the conclusion, $c, given$ the premises, is greater than the probability of the conclusion c alone. We now have our second condition. The induction "$p, \therefore c$" is a good one if

    ii)   the conditional probability of c given p is greater than the initial probability of c alone.

We have established two conditions which exist for both inductive and deductive logic. But what of a corresponding third condition? What, in induction, corresponds to the relation of entailment in deduction?

The part of logic that deals with deduction and entailment is a relatively settled and uncontroversial field. But the corresponding branch of inductive logic – the branch that deals with the logical relation between p and c when the *conditional* probability of c *given* p is greater than the *initial* probability of c alone – is ablaze with controversy! Decade by decade, the field of inductive logic approaches an impressive and more settled maturity; but it is not yet in a suitably composed state to be placed before newcomers. Thus our description of induction in this section stops with conditions i) and ii). The next section contains a very simple sketch of the mathematical basics of probability theory.

Chart I summarizes our description of inductive arguments, and contrasts their structure with the structure of deductive arguments.

Chart I

DEDUCTION	INDUCTION
A deduction is an argument in the form "$p, \therefore c$" where p is the premises and c the conclusion.	An induction is an argument in the form "$p, \therefore c$" where p is the premises and c the conclusion.
Each proposition in p has an initial truth-value; that is, it is either true or not true.	Each proposition in p has an initial probability, which is represented by a number between 0 and 1.
The premises together, p, will have a given truth-value, which depends on the truth-values of the individual premises.	The premises together, p, will have a given probability, which depends on the probability of the individual premises.

The argument p, $\therefore$ c is a good deduction when p verifies c.

The argument p, $\therefore$ c is a good induction when p increases the likelihood of c.

p verifies c when c is true on condition that p is true.

p increases the likelihood of c when the conditional probability of c given p is greater than the initial probability of c alone.

c is true on condition that p is true when p entails c.

The conditional probability of c given p is greater than the initial probability of c alone if ??? (The question marks indicate that a fully satisfactory answer has not yet been furnished by inductive logic.)

2. Inductive and Deductive Arguments

It is worth repeating that an argument can be examined from a deductive *and* an inductive point of view. In fact, one argument can be a good deduction and a bad induction. It can work the other way around, too. Consider the argument:

> Mercury is inhabited by intelligent beings.
> Venus is inhabited by intelligent beings.
> Jupiter is inhabited by intelligent beings.
> Saturn is inhabited by intelligent beings.
> Uranus is inhabited by intelligent beings.
> Neptune is inhabited by intelligent beings.
> Pluto is inhabited by intelligent beings.
> Earth is inhabited by intelligent beings.
> Therefore, Mars is inhabited by intelligent beings.

Obviously this is not a bad induction, but it is a bad deduction.[1]

Is it possible to have an argument that is both a good deduction *and* a good induction? This is not an easy question to answer, but it seems that it ought to be possible. Consider the following. Suppose that a team of social anthropologists has carefully studied an alien village in a distant land; the team now wants to prepare a classification of the basic social and functional rôles of the village. The anthropologist, Sue, investigates the written records in the basement of the Bureau of Records; Bill, her colleague, examines the records on the first floor. Sue makes a list of all males who have the status *lok-ko* ("bachelor"), and Bill compiles a list of all males who have the status *nahn-bownd* ("unmarried"). Sue and Bill have been working very hard, and Hank comes to give them a break. Though he does not know the language of the village, Hank is given the task of establishing co-ordination ratios between the categories on Sue's and Bill's lists.

As Hank proceeds, he notices the very impressive co-relation between *lok-kos* and *nahn-bownds*. Eventually, he offers this argument:

---

  (1)   Mr. X is both a *lok-ko* and *nahn-bownd*
  (2)   Mr. Y is both a *lok-ko* and *nahn-bownd*
  (3)   Mr. Z is both a *lok-ko* and *nahn-bownd*
  (4)   Mr. U is both a *lok-ko* and *nahn-bownd*
  (5)   Mr. V is both a *lok-ko* and *nahn-bownd*
  (6)   Mr. W is both a *lok-ko* and *nahn-bownd*
  (7)   Mr. A is both a *lok-ko* and *nahn-bownd*
  (8)   Mr. B is both a *lok-ko* and *nahn-bownd*
  (9)   Mr. C is both a *lok-ko* and *nahn-bownd*
(10)   Mr. D is both a *lok-ko* and *nahn-bownd*
(11)   Mr. E is both a *lok-ko* and *nahn-bownd*
(12)   Mr. F is both a *lok-ko* and *nahn-bownd*
(13)   Mr. G is both a *lok-ko* and *nahn-bownd*
(14)   Therefore, all *lok-ko*'s are *nahn-bownd*.

A good induction? It would certainly seem to be. But it is a good deduction as well. The statement, "all bachelors are unmarried," is true by definition. Since it is true by definition, it can not be false. A deduction is good when it is not possible for its premises to be true *and* its conclusion false. Since it is impossible for our conclusion to be false, it is impossible for its premises to be true and its conclusion to be false. Consequently, the argument is a good deduction.

Of course, an argument can also be hopelessly bad both as a deduction and as an induction.

An argument may be judged from quite different perspectives for different properties appropriate to those perspectives. What fails from one point of view may succeed from another, even though some arguments are failures from every logically respectable point of view.

In its most general form, an induction is any argument in which the premises increase the probability of the conclusion. The goodness of an induction is a matter of degree. In a deduction, p either entails c or it does not. But p might increase the probability of c only very slightly – or very considerably, or somewhere in between. Therefore, induction can be said to be slightly good or fairly good or very good indeed. A deduction, on the other hand, is an all-or-nothing argument: it is either a perfectly good deduction, or it is perfectly bad.

Induction is an utterly indispensable tool to generalize from the scattered and particular fragments of everyday experience – what the philosopher William James called "the buzzing, booming confusion". Every rule of thumb ("red sky at night, a sailor's delight"), every principle of the human personality ("frustration breeds aggression"), every principle of the market-place ("if demand is inelastic, an increase in supply will cause a larger fall in market price than an increase in amount sold"), every principle of human affairs ("men seldom make passes at girls who wear glasses"), every principle of experimental science ("copper conducts electricity"), involves the dynamics of induction. For we are speaking here not only of what is the case in *observed instances*, but also of what is the case in *unobserved* instances.

Induction, then, is the dynamic that underlies inferences from observed to unobserved cases. But it also structures *projections from past and present known cases to future cases*. If I say, on the basis of regularities already observed to exist, that copper conducts electricity, I speak for *then,* for *now* and for the *future*; I make a *projection*.

---

(1) A good induction is an argument in which the conditional probability of the conclusion, given the premises, is greater than the initial probability of the conclusion alone.

(2) Inductions enable us to "learn from experience", to enlarge our understanding of things by projections from the observed to the unobserved and from the past to the future.

We have seen that we can decide whether a given argument is deductively or inductively adequate. Here is an argument:

> All men are mortal
> Socrates is a man
> ∴Socrates is mortal

It is deductively correct, clearly. Its premises together *entail* its conclusion.

Here is another argument.

> The vast majority of men have a pineal gland
> Socrates is a man
> ∴Socrates has a pineal gland

Not, clearly, *deductively* correct. (Socrates *might* be a man without gland, though that is extremely unlikely.) Still, the argument is just as clearly a very strong *inductive* argument.

We might now conclude that we know at least something about what an argument is. At least we seem to know, definitively, that there are inductive and deductive arguments, and that we know how to identify them.

But we are not yet ready to make such a conclusion. Suppose that Sue advances an argument: "Most gorillas are vegetarians. Marvin is a gorilla; therefore, Marvin is a vegetarian." And suppose Bill counters: "This argument is deductively incorrect, therefore, it is a bad argument." Sue might well reply: "It is inductively strong; therefore, it is a good argument."

How are we to determine whether Sue's argument is a deductive or inductive argument? It is inductively strong but not deductively correct, but this is of no help. We know how it measures up against the standards of each category of evaluation, but that doesn't tell us which category it belongs to. We are back to one of our original problems: once we know that something is an argument, we can implement deductive or inductive methodology to tell us whether it is a good argument from each perspective, but we still don't know when something is an argument.

We might suggest that the argument is deductive if the person who advances it honestly intends it to be deductive. But an arguer's intentions cannot be determined by deductive or inductive logic. And people are at times very confused about what they honestly intend. We conclude, then, that inductive and deductive arguments cannot be located and identified by the purely inductive or deductive models of argument by appealing to the methods of these models.

It has been said that deductive arguments proceed from the general to the specific, whereas inductive arguments go from the specific to the general. But this is a misleading and incorrect account of the difference between deductive and inductive arguments. Some inductively strong arguments have a general premiss and a particular conclusion. For example: "All emeralds previously found have been green; therefore, the next emerald to be found will be green." And some deductively correct arguments proceed from particular premisses to general conclusions. For example: "The emerald numbered one is green, the emerald numbered two is green, and the emerald numbered three is green; therefore, all the emeralds numbered from one to three are green." There is no correlation at all between the nature of an argument – deductive or inductive – and the nature of the statements that comprise the argument – general or specific.

Summary

1. There are deductive and inductive arguments, but neither deductive

nor inductive logic can tell us how to distinguish between these two classes of argument.

2. Given any set of statements, the mechanisms of inductive and deductive logic will tell us that certain relations obtain among the statements; those relations determine whether a deductive or inductive argument is a good or a bad argument.
3. Given an argument, we can evaluate it from the perspective of the deductive or inductive model, whichever is appropriate.
4. Deductive and inductive logic operate within dialectic as components of a larger theory of argument. These components must always take the argument as given.

Given an inductive argument, what is it that makes it essentially an *argument*? To answer this question, we examine some elements of the theory of probability. We need a clearer picture of the inductive model of argument, so that we may compare and contrast it with the deductive model of the previous chapter.

3. Elements of Probability Theory

In that part of inductive logic known as probability theory, statements may be assigned a number between 0 and 1. The number is used to represent the degree of probability of any statement. For example, let us toss a fair die. Each of the six sides has an equal chance of coming up, and one of the six sides is numbered "three". Thus "A three will come up" has a probability value of $1/6$. At the extremities are the tautologies of deductive logic – "either a three will come up or not". The extremities have the value 1. The contradictions of deductive logic ("a three will come up; a three will not come up") have the value 0. The cases we are mainly interested in, the cases of inductive logic, will have a value represented by a fraction somewhere between 0 and 1.

For each assessment, we can say, "The probability of (statement) p is x (a value expressed as a number)." This assessment is expressed as a formula: "$\text{Pr}(p) = x$". We have formulated the probability of the statement, "a three will come up," as $1/6$. By using our formula, we can now formulate *negative* probabilities.

Negation: $\text{Pr}(\daleth\, p) = 1 - \text{Pr}(p)$

For example, the probability of not getting a three on a throw of a fair die is

$1 - \text{Pr}$ (a three is thrown)
$= 1 - 1/6 = 5/6$.

We can also evaluate disjunctions. If two statements are mutually exclusive – that is, if it is not possible for both to be true at the same time – then they are said to be disjunctions. To calculate their disjunctive probability, determine the probability value of each statement; add the two values. The sum is the disjunctive probability.

Disjunction: if p and q are mutually exclusive, then

$\Pr(p \lor q) = \Pr(p) + \Pr(q)$.

For example: we may assume that it is not possible for a fair die to show two faces uppermost on one throw. Hence "a six will come up" and "a three will come up" may be treated as mutually exclusive statements. Thus Pr (a six will come up) + Pr (a three will come up) = $^1/_6 + {}^1/_6 = {}^1/_3$.

However, it is possible to formulate two statements that are not mutually exclusive. We need a formula that will express the possibility of their *joint* truth; in other words, the *conjunctive* probability of the statements must be taken into account. A more general approach to disjunction will help us.

Disjunction: $\Pr(p \lor q) = \Pr(p) + \Pr(q) - \Pr(p \land q)$

To illustrate this equation, we will consider the chances of getting either an odd number or a number greater than 3. Pr (an odd number will come up) = $^3/_6 = {}^1/_2$; Pr (a number greater than 3 will come up) = $^3/_6 = {}^1/_2$. We might think that the earlier rule – add the two fractions – would give us the disjunctive probability. But adding the fractions yields a disjunctive probability of 1. This result is incorrect, however – there is the possibility of getting an even number less than 3. For example, we might roll the die and see a 2 – which is neither odd nor greater than 3. This possibility is expressed by the equation "Pr ⅂ $(p \lor q)$". The probability of neither p nor q is, by the negation rule, greater than 0. Thus we can see that Pr $(p \lor q)$ must be less than 1. In adding the two fractions together, we would have committed the fallacy of mutual exclusiveness: we did not account for the non-mutually-exclusive character of the disjuncts. It is possible that we might get a toss that represents both an odd number and a number greater than 3; the probability of this eventuality is $^1/_6$. To solve the equation Pr $(p \lor q)$, calculate Pr (an odd number will come up) + Pr (a number greater than 3 will come up); subtract Pr (an odd number greater than 3 will come up). The result is $^1/_2 + {}^1/_2 - {}^1/_6 = {}^5/_6$.

Conjunction

The conjunctive probability of two statements, p and q, may be defined as their product provided the two statements are probabilistically independent of each other.

> Conjunction: Provided p and q are probabilistically independent, $\Pr(p \land q) = \Pr(p) \times \Pr(q)$.

> Assume you have two dice; the face shown on one does not influence the face shown on the other. The probability that both will come up 3 is calculated by multiplying together the probabilities that each will come up 3. That is, $^1/_6 \times {}^1/_6 = {}^1/_{36}$. Generally, we can see that conjunctive statements tend to diminish in probability values as more conjuncts are added.

It is important to understand the difference between *probabilistic independence* and *mutual exclusiveness*. The statements, "Greta Garbo ate a Big Mac in Portage La Prairie on June 8, 1979" and "This die will come up three" may be presumed to be probabilistically independent of each other. But they are not mutually exclusive, because it is quite possible that both might be true. On the other hand, the statements "This die will come up three" and "This die will come up four on the same toss" are mutually exclusive, but they are by no means independent. The outcome of one statement affects the probability of the other. Indeed, if one statement is true, the probability of the other is zero. Probabilistic independence may be defined in terms of conditional probability:

> Independence: p and q are independent where $\Pr(q$ given $p) = \Pr(q)$.

It is easy to overlook probabilistic dependencies in estimating conjunctive probabilities. This mistake is called the *fallacy of independence*. As an example, let us calculate the chances of drawing two red cards from a deck of fifty-two cards. It might seem logical to multiply $^1/_2$ by $^1/_2$, since half the cards in the deck are red. But if we do not replace the first card in the deck before the second is drawn, our equation is fallacious. Unless we replace the first card, the second draw is not independent of the first. At the time of the second draw, one card has been removed from the deck; this affects the probabilities of selecting a red card on the second draw.

Another fallacy common in conjunctive inductive reasoning is the fallacy of *confusing prior and posterior probabilities*. Suppose that we have drawn a red card from a deck of fifty-two cards; suppose, also, that we wish to calculate the probabilities of drawing two red cards. Assume that the card we have drawn is replaced in the deck, and that the deck is reshuffled. Again, we might be tempted to think that the probability of drawing two red cards is $1/2 \times 1/2 = 1/4$. But keep in mind that *we have already drawn one red card.* Our original statement would have been something like" I will draw one red card from this deck; I will draw a second red card from this deck." The first part of the statement has come true; thus it *now* has a probability factor of 1; its outcome is known. Once one card has been drawn, we need a new equation. An accurate calculation would be: $1 \times 1/2 = 1/2$. In other words, if one red card is already drawn, we can base our calculations on the second draw only. It is, after all, the only remaining unknown outcome; it now may be treated as independent of the previous draw. We need to bring our data up to date.

Stephen K. Campbell describes some interesting examples of *confusing prior and posterior probabilities.*[2]

He tells the story about the man who packed a harmless bomb in his luggage for protection on plane trips. The man reasoned that the chances against *two* people carrying a bomb on the same plane were astronomically high. The chances of one person carrying a bomb are very low; thus the chances of two people doing it would be extremely low. However, as Campbell points out, the man who takes the bomb along knows with certainty that he has it with him. If we take this knowledge into account, we can hardly say that the chances of one person carrying a bomb are low. Hence we cannot say that, by carrying a bomb, the man reduces the probabilities of a second bomb on the plane.

Another example: a sailor stuck his head through a hole that a cannon-shot made in the side of a ship. He reasoned that it was unlikely another cannonball would come exactly through the very same hole.

Conditional Probability

Conditional probability is expressed in terms of conjunctive probability:

$$\Pr(q \text{ given } p) = \frac{\Pr(p \wedge q)}{\Pr(p)}$$

Notice that the "definitions" of negation, conjunction, disjunction, and condition cannot be reduced to some one basic notion, which can be calculated independently. But the "definitions" we have given are characteristic of the relations that form the structure of inductive arguments. They *partially* illustrate how inductive arguments work; they partially illustrate how the logical relations between statements in inductive contexts differ from the logical connections between statements in deductive logic.

We might even be tempted to think that there is a certain *circularity* implicit in our sequence of "definitions". For example, we have defined conditional probability in terms of conjunctive probability. This will cause a problem if we want to calculate conditional (q given p) probabilities where Pr (p) is not independent of Pr (q). If p and q are independent, then Pr (q given p) is simply Pr (q). But if Pr (p) is not independent of Pr (q), then our rule for conjunction *cannot* apply; that rule requires the independence of p from q. What we need is a more general conjunctive rule that does not presuppose independence.

Conjunction. Pr ($p \wedge q$) = Pr (p) $\times$ Pr (q given p)

There is a problem: this rule presupposes that we already know Pr (q given p). But Pr (q given p) is a conditional probability – and it was conditional probability we set out to determine in the first place! Thus Pr (q given p) cannot always be determined by looking exclusively at the components, Pr (p) and Pr (q). Here, then, is a significant difference between inductive logic and classical deductive logic, where the truth-value of $p \supset q$ is determined solely by the truth-values of p and q.

So far as we have here developed the theory of probability, initial probabilities do *not* uniquely determine conditional, conjunctive or disjunctive probabilities in many cases.

Some idea of the shape of inductive logic does emerge through the comparison of *inductive* negation, conjunction, disjunction and conditional probability with their *deductive* counterparts. Certainly we can now better appreciate the distinction between initial probabilities and conditional probabilities.[3]

Exercise

Given a fair deck of fifty-two playing cards, what is the probability of drawing each of these hands: (a) an ace; (b) a card that is not an ace; (c) a black ace; (d) an ace or a spade; (e) an ace or a non-spade; (f) a black ace or a card that is not an ace; replacing the first card drawn, (g) an ace on the first draw and then, after replacing the first card drawn, a spade on the second draw; (h) two aces; (i) an ace and a jack; (j) an ace or a jack.

4. *Secundum Quid* or Hasty Generalization

One of the most widespread and pervasive inductive fallacies of everyday reasoning is over-generalization (or, as it is sometimes called, "hasty generalization" or "glittering generality"). How often does "popular wisdom" – "All women are fickle"; "All Germans are methodical" – go unchallenged in the face of obvious evidence to the contrary?

Antoine Arnauld offered some highly characteristic types of unwarranted inductions, in which a general principle is based on the evidence of a few particular observations.[4]

> There are diseases which escape the detection of the most skilled physicians, and prescribed cures are not always successful.
> ∴ Medicine is completely useless and is a craft of charlatans.
>
> Great vices are often concealed beneath a facade of piety.
> ∴ All devotion is hypocrisy.
>
> Some things are difficult, and often we are grossly deceived.
> ∴ We cannot know the truth of anything.

Traditionally, this sort of fallacy is called *secundum quid*, "in a certain respect". What is true in a certain respect need not be true in all relevant respects. Fallacies of *secundum quid* neglect some of those relevant respects. There are problems inherent in implementing a working decision procedure for the *secundum quid* fallacy. There seems to be no general method available for telling us exactly when all the necessary qualifications to a statement have been made, or when the loopholes in a generalization have been plugged. However, the branch of probability and statistics known as *sampling theory* sheds some light on the problem.

What is a sampling procedure? Consider an urn containing some marbles. We know that some of the marbles are black and some white. The best way to find out how many marbles are black and how many are white would be to remove the marbles and count how many of each colour there are. Sometimes, however, it might be inconvenient or even impossible to do this. If there are *very many* marbles in this particular urn, then we might have to make a conjecture about the proportion of black to white marbles without actually counting them.

In real-life generalizations, access to data is rarely complete. Fortunately, however, there is usually an *indirect* method of obtaining access to data. Let us go back to our urn. We could pull out a handful of marbles and count the proportion of black to white in the handful. We would be working on the assumption that the marbles in the urn are mixed so that the proportion of black to white is approximately the same throughout. If this assumption is reasonable, if the sample is large enough to be meaningful, and if it is selected properly, we can get a good idea of the composition of the whole "population" of marbles in the urn.

These are big ifs, to be sure. (1) If we picked just one marble, and it were black, should we conclude that *all* the marbles are black? Such a generalization would be *virtually meaningless*; the sample is not large enough to allow us to conclude much at all. (2) If we picked a large enough sample but purposely selected only black marbles, ignoring the white ones, we might conclude that all the marbles are black. But our conclusion would be *biased* and *misleading*. (3) Suppose we happen to know that all the black marbles are at the top, and all the white ones at the bottom; suppose we take our sample from the top only. It would be misleading to generalize from the composition of the sample to the composition of the whole population of the urn. (4) Generally, *any sampling procedure is effective only where the sample is representative of the whole population in respect to the composition and distribution of the property in question throughout that population*. The traditional fallacy of *secundum quid* becomes more clear in the context of sampling theory. Actually, two distinct types of fallacy are involved.

One is the fallacy of *insufficient statistics*. This occurs when the sample chosen is so small that to generalize from the sample to the entire population is virtually meaningless. In everyday argument, the basis of a generalization is quite often slim when considered thoughtfully and critically. We are told, for example, that a given "test group" had X per cent fewer cavities. But the size of the "test group" is never specified! If the test group were small enough, any generalization about the entire population based on the test group would be worthless. Often phrases like "according to a scientific study" or "evidence suggests" are simply a cover for a complete lack of evidence.

A sample must be sufficiently large to constitute *seriously worthwhile evidence* for a generalization to a given population – particularly if that population is a difficult or tricky one to generalize about. Too small a sample can be radically misleading, especially where the population is variegated.

Even if the sample is large enough, we can still find the fallacy of *biased statistics*. This fallacy occurs when a generalization is based on a sample that does not represent the population in one particular way – that is, the sample does not tell us enough about the *factor* we are attempting to analyze. In our urn illustration, we saw that if we picked only black marbles for our sample, that sample would not tell us much about *all* the marbles in the urn. Or if the black marbles were all at the top of the urn and we only took marbles from the top, our sample would not be representative of the colour of the marbles in the whole population. These are *biased* samples. If there is a variegation in the population, the sample must reflect those varieties. For example, suppose that it is known that we had a layer of black marbles, then a layer of blue, red and white marbles respectively. Clearly, a meaningful sample could be selected only by taking an equal number from each layer (assuming each layer is of equal size). To deal with possible variegations, our sampling procedure must be *representative* of the composition of the population.

The simplest kind of fair sample is called the *random sample*. In a random sample, each variegation in the population *must have an equal chance to be in the sample*. In any kind of practical poll, survey, or generalization, however, it is rarely possible to proceed by picking a random sample. There are usually so many known variegations in a population, that the likelihood of bias would be great.

Where there is regular variation in the population, a *stratified random sample* is required. A random sample is taken from each *stratum*. For a stratified random sample, the population must be classified into mutually exclusive sub-groups (strata); then independent random samples must be drawn from each sub-group. Thus, in our example of the urn, it would make sense to take an equal number of marbles from each layer, assuming the layers are of equal size.

A third technique is *cluster sampling*. A population is divided into groups called "clusters"; then a sampling of the clusters is taken. For example, if the population to be studied consisted of all the households in Canada, we might take a random sample in Canada and then select a random sample of households in each area.

The fallacy of inadequate statistics and the fallacy of biased statistics are two sides of the same counterfeit coin. A very small sample, in proportion to a total population, is much more likely to be biased than a very large one. If the sample is very small then there is a very great risk of error; if the sample is small enough, then any statistical generalization derived from it may be virtually worthless. Despite its inaccuracy, such a small sample might still impress an audience.

It is a mistake to infer that what is true of a large sample must be true of a smaller sample. The example of penny-flipping might help to illustrate two fallacies that can occur by making such an inference. First: statistical theory tells us that if a penny is a "fair toss" – if heads and tails are equally probable on each throw – then, given a *large* number of tosses, the chances are reasonably good of getting a roughly equal proportion of heads and tails. But it does not follow that the chances are *also* reasonably good of getting half heads and half tails in a small number of tosses.

Second: a "fair toss" means that each outcome in a sequence is independent from its predecessors or successors in that sequence. That is, if you got heads on this toss, it does not follow that it is more likely you will get tails (or heads for that matter) on the next toss. Each toss is independent of the previous one. But we do tend to regard such sequences as "self-correcting". For example, given a run of tails, we expect that the next toss is more likely to be heads. This is a fallacy. It is a common one, with a recognized name, the *gambler's fallacy*.

In part, the gambler's fallacy consists of confusing a single member of a sequence with the whole sequence. Statistics tells us that a sequence of a large number of heads, H H H H H H H, is a slightly less probable outcome than a sequence that contains a tail, H H H H H H T. Instead of thinking of the whole group of outcomes, though, we will concentrate on just one throw. The chances of getting heads are *just the same* as the chances of getting tails: $1/2$. For each single toss, we must disregard what went before.

Applying the notions of conditional and conjunctive probability, we can see that the gambler's fallacy has a *dual aspect*. First, the gambler expects that, given a certain outcome on one toss or a series of tosses, it is more likely he will get a certain outcome on the next toss. He believes that Pr (tails will come up on this toss, given that heads came up on the last toss) > Pr (tails will come up on this toss). He believes that the conditional probability of one statement (q, given p) is greater than the non-conditional probability of q taken by itself: Pr (q given p) > Pr (q). His belief is a denial that p and q are independent.

We said that p and q are probabilistically independent only where Pr (q given p) = Pr (q). But it is reasonable to assume that each toss must be independent of the next. Thus, if the tosses really are independent, the gambler commits a fallacy!

The fallacy consists of a logical inconsistency, which concerns independence. In theory, the gambler may well concede that each toss is independent from the next. But it may be hard for him to adhere consistently to the assumption of independence, given his *psychological* expectation that sequences of events tend to be "self-correcting". The fallacy is a failure to understand the *idea* and *application* of conditional probability.

But assume we find a gambler who consistently adheres to the assumption of the independence of each toss in the sequence. That gambler may be committing a second, and different, fallacy. Remember that the gambler thinks of the whole group of outcomes instead of concentrating on just one throw. If he adheres to the assumption of the independence of each toss then, by our rule of conjunction, he can calculate the *probability of a sequence* by multiplying the individual probabilities together. For example, the probability of getting three heads in a row would be $1/2 \times 1/2 \times 1/2 = 1/8$. The *conjunctive* probability for the three tosses is Pr ($H_1 \wedge H_2 \wedge H_3$) = $1/8$. But this does *not* mean that Pr (H_3) = $1/8$! The probability of *each individual* toss coming up heads, assuming independence, is $1/2$. The gambler confuses Pr ($H_1 \wedge H_2 \wedge H_3$) with Pr ($H_1$) $\wedge$ Pr (H_2) $\wedge$ Pr (H_3). He forgets that he is not betting on all three tosses, but only on the one. It is critical to distinguish clearly between the *conjunctive probability* of a *sequence* of outcomes and the *individual probability* of *each outcome in the sequence*.

What is the psychology of inadequate statistics and biased statistics? Why is it so easy to commit these fallacies? Why are they so widespread? According to some writers, many of us have a strong tendency to think that samples of a population are more representative of that population than sampling theory would indicate.[5] When subjects are instructed to predict the proportion of heads in a random sequence of fair tosses, for example, they usually produce sequences that are far closer to ½ than the laws of chance would allow for short sequences. That is, subjects tend to have unrealistically high expectations about the "fairness" of the coin, especially for short sequences of tosses.

Another general tendency is to think that short sequences of tosses are "self-correcting". For example, many people think that, after a string of heads in a sequence of tosses, there should be a trend towards tails. Tversky and Kahneman suggest that this phenomenon is the heart of the gambler's fallacy:

> The gambler feels that the fairness of the coin entitles him to expect that any deviation in one direction will soon be cancelled by a corresponding deviation in the other. Even the fairest of coins, however, given the limitations of its memory and moral sense, cannot be as fair as the gambler expects it to be.[6]

The fallacy would seem to consist, then, of the illegitimate inference that what is true of large numbers of things must also be true of small numbers of things. Very large numbers can, under the right conditions, be highly representative of the population from which they are drawn. But it is incorrect to expect that what we learn from the large numbers can be applied to small numbers as well.

The psychology of small numbers suggests that biased and inadequate statistics are occasional lapses of the unwary; they also represent forms of inference that are virtually irresistible in everyday argumentation and even in scientific research. Tversky and Kahneman suggest that, too often, psychological studies are based on ridiculously small samples. The small samples can result in frustrated scientists and inefficient research. Small wonder, then, that the fallacy of inadequate statistics is so prevalent in the less structured milieu of everyday reasoning and decision making.

5. Another Deceptive Duo: Meaningless and Unknowable Statistics

Daniel Seligman has called one fallacy the fallacy of *meaningless statistics.*[7] This fallacy occurs when the terms used in a statistical claim are so vaguely defined that the statistic is virtually meaningless. He cites, as an illustration, a statement attributed to Robert Kennedy: "Ninety per cent of the major racketeers would be out of business by the end of the year if the ordinary citizen, the businessman, the union official, and the public authority stood up to be counted and refused to be corrupted." The sentiment is admirable, but unless there is some general agreement on what a "major racketeer" is, or on what constitutes standing up to be counted, it is absurd to claim that an exact ninety per cent of anything can be dealt with.

If we thought about it, it is not likely that we would take the exact figure of ninety per cent too seriously anyway; perhaps we weren't meant to. But if we didn't stop to think about it, we would quite probably be favourably impressed by the precision of "ninety per cent". Unless there is some agreement on precise and objective criteria, it is silly to try to relate exact statistical figures to many of the terms used in ordinary, common-sense language. Such terms are often extremely vague. The definition of "major racketeer" will likely vary from New York to Calgary to Treasure Island.

In some ways, then, the fallacy of meaningless statistics is a *linguistic* fallacy. It violates the requirement that our terms must be defined precisely enough to make exact statistics meaningful. We are swamped with figures on crime and unemployment, to take two prominent examples, but we are rarely told how terms such as "unemployed person" are being used.

A second fallacy is the fallacy of *unknowable statistics*. The meaning of the terms used by the arguer might be adequately clear, but facts he alleges are ones that could not possibly be determined. Seligman cites a claim made by Dr. Joyce Brothers: "The American girl kisses an average of seventy-nine men before getting married." Even if enough American girls had such idiosyncratic memories, it is extremely implausible to think that they might produce such accurate tallies. So we need to ask: how could one obtain such information?

An especially popular example of unknowable statistics is found in the figures that are given in newspapers concerning the rat population of New York City. Seligman reports that, for years, newspaper feature writers have claimed that there are eight million rats in New York City. This sounds impressive until one thinks about some of the difficulties inherent in making a reasonable estimate of New York's rat population. The rats themselves do not tend to be very co-operative. Seligman interviewed the Rodent and Insect Consultant for the City of New York, and was referred to two studies of the problem. The investigators counted rats in certain areas; then, from these calculations, they extrapolated estimates for the entire city. However, the estimates exhibited considerable variation. This variation is understandable when one considers the problems confronting the investigators. As the Insect and Rodent Consultant put it: "You can count a rat on the eighth floor of a building and then another on the seventh floor, and then another when you get to the sixth – but after all, you may just be seeing the same rat three times."

Meaningless statistics and unknowable statistics are dangerous and important pitfalls of statistical reasoning and argument. But neither fallacy can be studied adequately by standard methods of probability and statistics unless they are also examined from the wider perspective of the argument. As we have found, the fallacy of meaningless statistics is essentially a *linguistic fallacy*. The fallacy is committed when we ignore the standards of precision that are specified by linguistic conventions. These conventions are different for each different context of argument, and standards of precision will vary as well. An argument between experts on a theoretically highly developed technical subject will have different standards than an argument between laymen on a subject that may be difficult to quantify.

The fallacy of unknowable statistics pertains to those logical or practical impossibilities that may be inherent in collecting evidence.

Such impossibilities are not confined to purely statistical arguments. They crop up whenever there is reference to evidence that is intrinsically unobtainable. We might suspect that this fallacy is present if it were argued that trees grow because they are inhabited by invisible and otherwise unobservable spirits. If the spirits are unobservable, there would seem to be little possibility of refuting – or confirming – their existence. What is missing in such arguments is *accessibility to evidence*.

6. Statistical Graphs

Charts, graphs and diagrams that organize and summarize data are frequently found wherever statistics are used. But these pictorial representations can be especially deceptive. They admit of unscrupulous use; there are various elementary tricks associated with their use. It is good to be aware of such tricks. Suppose, for example, that we have some set of figures that can be indexed on a scale from 1 to 10; suppose that those figures show an even rise from 7 to 8 between 1963 and 1978. The figures could represent profits per average shareholder in a company. Graph I tells us that, between 1968 and 1978, shareholders' profits went from $7,000 to $8,000. Not bad for the shareholders.

GRAPH I

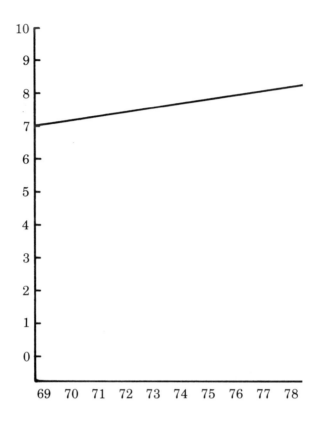

An easy way to make the picture look better *without altering the figures* is to chop off the blank spaces. Graph II looks even more impressive.

GRAPH II

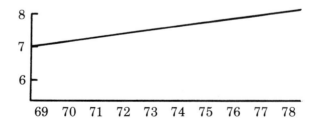

Better yet, let's change the scale of the column of figures. We will make the chart represent only the interval between 7 and 8; the increments of the scale are also changed. Our results are shown in Graph III.

Graph III

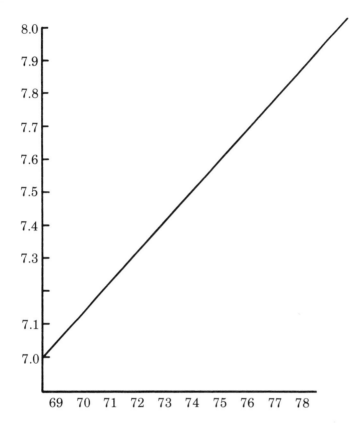

Graph III looks very impressive indeed. Our shareholders' profits now give the appearance of shooting up at a terrific rate. But notice that we need not tamper with the *statistics* to give this impression of dramatic rate of increase. All we have done is manipulate the scope and scale of the graph. We have made it look as if profits are skyrocketing, although the graph indicates only what it showed before – that profits per shareholder have gone up from $7,000 to $8,000 in the ten-year period, 1968-78.

This widespread technique is dishonest. But it is also universal – it can be applied to any graph. Given a plotting of any two sets of figures on a graph, a more dramatic picture can be created merely by *chopping* the blank space or by changing the graph's *intervals*.

One can alter the picture even more by *stretching* or *shrinking* the scales of the intervals that appear on the graph. To make a graph's curve rocket skyward in a spectacular trajectory, we can *shrink* the horizontal intervals and *stretch* the vertical ones. This technique is illustrated in Graph IV.

GRAPH IV

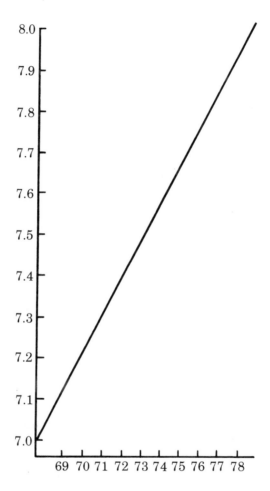

The figures are still the same. But the shareholders' profits are shooting upwards *most* dramatically!

If your audience is particularly obtuse, you can resort to the most underhanded tactic of all. Simply *remove* the interval scales altogether, and draw the curve however you like. Our shareholders' profits are shown again in Graph V.

GRAPH V

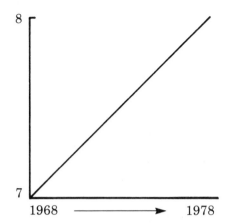

1968   ⟶   1978

But Graph V looks a little bare. For your presentation, you could dress it up with some pictures or cartoons. For example, since it is intended to illustrate a rise in profits, you could try something like Graph VI.

GRAPH VI

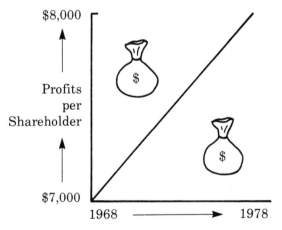

Graph VI represents an interesting strategy of deception, which might be called the technique of *erasure*. We have removed the figures from the intervals of the graph. This strategy is based on the hope that the uncritical reader will not look for *figures*, but will be impressed by what appears to be an upward trend. Stephen K. Campbell gives sound advice when he writes, "Charts with unmarked axes should never be trusted.

When the creator of a chart is free from the discipline imposed by well-marked areas, he can convey through the chart just about any impression he chooses."[8] Unless the intervals of measurement are clearly marked out on the axes, the graph should not be taken as a serious attempt to convey reliable information. An unmarked chart allows the persuader to interpret his statistics in whatever way might seem most favourable to his argument.

Variations on this theme are well known, and bar-graphs, pictograms, and other suggestive yet conceptually fuzzy forms of illustration often find their way into statistical presentations. Most of us don't want to take time to digest a welter of complex numerical tabulations; we are quite happy to see statistical information neatly summed up in an appealing picture. Such pictures have immediately evident "conclusions". But, if you are interested in *reasoned argument*, a healthy distrust of pretty pictures is not a bad attitude to foster. The attractively packaged conclusion may be quite fallacious, just as the attractively packaged product it was designed to promote may be a bad buy.[9]

Exercise

1. Write a report that analyzes an article from the media. Choose an article that makes general statistical claims, and comment on the strengths and weaknesses of the article's argument.
2. Find an example of a survey or poll and study its sampling procedures. Try to evaluate whether statistical fallacies could be committed.

Summary: An Economist's Point of View

We end our discussion of statistical graphs with a useful quotation from Leonard Silk.

> Graphs and charts can easily be drawn in such a way as to mislead the reader; small changes in numbers can be blown up, or large changes shrunk, by changes in the vertical or horizontal scale of a chart, by not using a zero base, by the misuse of bars or drawings, etc. Statistics *don't* speak for themselves; like words, they can be used honestly or dishonestly.[10]

7. *Post Hoc:* A Causal Fallacy

Post hoc ergo propter hoc means "after this therefore because of this". The term is usually shortened to *post hoc*. The *post hoc* fallacy occurs when we observe that one event, B, follows another event, A, in temporal sequence, and then hastily conclude that A causes B. For example, one is often tempted to conclude, from the downpour that usually happens after one washes the family car, that somehow the rain and car-washing must be causally related.

Post hoc is sometimes described as the fallacy of arguing from correlation to causation, but the fallacy must be carefully described. A correlation between two classes of events is often evidence that they are causally related – but it is not conclusive evidence. There is nothing wrong *per se* with the following kind of argument: there is a correlation between events of kind A and events of kind B; therefore A causes B. Rather, the fallacy occurs when we leap too quickly to a causal conclusion from a premiss of observed correlations. In so arguing, we may be overlooking certain factors. Actually, there are a number of factors that get overlooked; consequently there are various types of *post hoc* fallacy.

8. Seven Sophisms

Within traditional treatments of the fallacy of *post hoc*, it is possible to distinguish seven distinct causal fallacies. We will outline the main thrust of each of these errors. For each error, we will give a characteristic example in an attempt to see exactly what is wrong with that sort of argument.

(1) Concluding that B was caused by A just because B happened after A.

Example:

> I took a dose of Sinus Blast, and a couple of days later my cold cleared right up.

The suggested conclusion, that taking Sinus Blast caused the cold to disappear, is fallacious. Perhaps the cold would have cleared up by itself after a couple of days, even sooner, if Sinus Blast had not been taken. Another stock example concerns the behaviour of a passenger on board the doomed Italian liner, *Andrea Doria*.

> On the fatal night of the *Doria*'s collision with the Swedish ship, *Grisholm*, off Nantucket in 1956, the lady retired to her cabin and flicked the light switch. Suddenly there was a great crash and the sound of grinding metal. Passengers and crew ran screaming through the passageways. The lady burst from her cabin and explained to the first person in sight that she must have set the ship's emergency brake.[11]

To assume a causal connection from a *single instance* of a given simple sequence of events presents a great risk of fallacy. Single instances have a way of being coincidences.

(2) **Concluding that B was caused by A just because there was a positive correlation between some previous instances of A and some previous instances of B.**

Example:

> Near perfect correlations exist between the death rate in Hyderbad, India from 1911 to 1919 and variations in the membership of the International Association of Machinists during the same period.[12]

In the absence of further evidence of a causal connection between these remote sets of data, to conclude that one is the cause of the other would be absurd. In certain circumstances, a high positive correlation may be good evidence of a causal link. But in this case, "common sense" indicates that it is extremely unlikely that the two sets of phenomena are causally connected.

(3) **Reversing cause and effect.**

Example:

> The people of New Hebrides have observed, perfectly accurately, over the centuries, that people in good health have body lice and people not in good health do not. They conclude that lice make a man healthy.[13]

Lice were normal among these people. A little further observation reveals that when anyone took a fever and his body became too warm for comfortable habitation, the lice departed. Thus there is a causal relation between the lice and good health, but the conclusion of the New Hebrides people reverses cause and effect. The causal relation is non-symmetrical, and an additional inference is required to establish which correlate might be the cause of the other. Quite often a genuine causal relation exists where it is not evident which of a pair of correlates is the cause and which the effect. For example, look at the correlation between income and ownership of stocks.

(4) **Concluding that A is the cause of C when both are the effects of a third factor, B.**

A correlation between A and C may not be indicative of a causal relation between A and C:

Instead, there might be some third factor, B, which causes both A and C. Factor B thus accounts for the correlation.

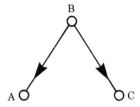

To assume a connection between A and C is wrong; such an assumption is called a spurious correlation. Compare the spurious correlation with a deceptively similar situation, where B is an *intervening variable* between A and C:

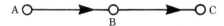

The intervening variable allows a genuine causal relation to obtain between A and C. In both cases, there is a relationship between B and C:

The defining feature of spurious correlation is that C does not cause B, although B causes C. Two examples will make the contrast clear.[14]

Example 1: In a survey on factory absenteeism, it was found that married women had a higher rate of absenteeism than single women. Later, it was found that the absenteeism rate among married women was almost as low as the absenteeism rate of single women if both groups had little or no housework; absenteeism among single women was almost as great as absenteeism among married women if the single women had a great deal of housework. A further survey showed, as one would expect, that married women generally had more housework than single women. We could correctly conclude that getting married causes more absenteeism. But a more complete explanation may be found if we include the intervening factor of housework: getting married causes more housework causes more absenteeism. Here we have a genuine causal relation mediated by a third factor.

Example 2: It was found that married persons ate less candy than single persons. A second look at the data showed that if married and single persons of equal age were compared, the correlation between marital status and candy consumption disappeared. It would be misleading and incorrect to conclude that getting married causes less candy eating. Age is the operative factor in both increased likelihood of marriage and decreased candy consumption. Thus:

There is a critical distinction between this example and the housework example. Here, it is incorrect to conclude that getting married causes getting older. If we could prevent people from getting married, we would not have stopped them from getting older. But the converse causal relation might obtain: if we could stop people from getting older, we might effectively reduce the incidence of marriage.

As we have seen, intermediate causation can be represented graphically:

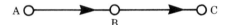

In intermediate causation, a fallacy can arise. We might omit intermediate factors – or add them – or we might not characterize correctly the nature of the causal relation between A and C. Take the case of Mr. X. While driving in traffic, X repeatedly observes that whenever he applies the brakes the defroster fan squeaks. He concludes that the brakes are connected to the fan. A more mechanically sophisticated observer might infer that applying the brakes causes deceleration of the vehicle, which tilts the fan mechanism, which in turn causes the squeak.

Let us return to the candy and housework examples. The final test of the merit of the distinction between the two cases is essentially a *practical* matter. A factory manager might institute the policy of discouraging female employees from marrying; this policy would effectively tend to reduce absenteeism. Remaining single means less housework and less housework means less absenteeism. But suppose a candy manufacturer were to succeed in discouraging large numbers of people from marrying. Would this lead to an increase of candy consumption? Clearly not. To increase candy consumption, he would have to keep people younger. Preventing them from marrying would not keep them from getting older. The distinction between causation and spurious correlation rests on a practical question: is A a *practically* necessary condition of C? If I stop A, will C stop, too? Or can I only stop A and C through a third factor, B?

The fallacy arises when we confuse two kinds of causality, the intervening variable and the antecedent mutual case.

(1)   *intervening variable*   (2)   *antecedent mutual cause*

When we confuse (1) and (2), we confuse the correct inference, (1*), with the incorrect inference, (2*).

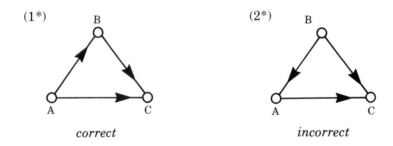

(1*)   *correct*   (2*)   *incorrect*

Provided the causal relation is transitive, (1*) is correct. To say that the causal relation is transitive is to say that if A causes B and B causes C, then A causes C. For example: if smoking causes emphysema and emphysema causes unhealthy lung obstructions, then smoking causes unhealthy lung obstructions. The second example, (2*), is incorrect. Though fallacious, (2*) may seem correct if we assume that the causal relation is symmetrical. The assumption of symmetry involves an inference:

> B causes A.
> Therefore, A causes B.

This illicit assumption may create the illusion that (2*) and (1*) are the same. But there is a difference! What distinguishes the fallacious (2*) from the correct (1*)? The statement A o———→———o B fails in (2*); in (1*), that relationship exists.

(5) Confusing causation and resemblance.

David Hackett Fischer offers a vivid example of this fallacy.

> The Picts constructed brochs and souterrains that were small, dark, and mysterious. Therefore the Picts themselves were small, dark, and mysterious.[15]

The fallacy seems to form the basis of many traditional folk remedies, for example, the belief that a bloodstone will cause bleeding to stop.

(6) Citing a pragmatically otiose necessary condition.

Example:

> Smith drowned because he did not learn to swim when he was young.

The causal attribution strikes the ear as far-fetched. More immediate factors, would be of more help in explaining Smith's demise, factors more at hand to possible reversal and more proximately previous to the drowning. However, in the context of a coach's presentation of the benefits of swimming lessons, the causal claim in the example might not appear quite so remote. Again, causal "talk" is strongly linked to practical production and prevention. Thus, it can be true that A is genuinely a necessary condition of B, yet it can still be fallacious to conclude that A caused B when control of A is impractical, irrelevant, or impossible.

(7) Overlooking or suppressing information that may run counter to the apparent trend of the correlation.

Example:

> It is easy to show by figures that the more it rains in an area, the taller and better the wheat grows. Conclusion: rain is good for the crops.

In this example, the positive correlation holds – up to a point. Beyond that point, the correlation takes on a negative significance. The fallacy is similar to the familiar abuse of sampling theory, in which an unrepresentative sample is selected. An instance of $A \wedge \neg B$ can overturn a causal allegation of $A \supset B$; negative correlations must also be considered. An allegation of "If there is war there is famine" can be overturned by an instance of war without famine. Or, as Mill would have put it, the method of agreement needs to be supplemented by the method of difference.

9. Analysis of *Post Hoc*

It is not easy to offer an analysis of the fallacy of *post hoc*, because there is no widespread agreement on how to analyze the concept of causation. Consequently, no established theory of what, precisely, is wrong about *post hoc* reasoning can be offered at present. It is clear that *post hoc* is an inductive fallacy of some sort, but of exactly what sort? At present, the answer is beyond the limits of inductive logic. We seem to understand intuitively what causation is at a practical level. But numerous philosophical critiques of the concept show that it is highly elusive and unstable at the level of theoretical analysis.

We have seen that *post hoc* is really an umbrella term for seven different kinds of fallacies, each of which has something to do with causal reasoning. And that is about as far as we can go in offering an account of *post hoc* without entering into an analysis of causation.[16] We conclude our discussion with a negative finding: whatever causation is, we must be aware of the numerous pitfalls we might encounter if we conclude too quickly that A causes B just because there is a correlation of some sort between A and B. The danger of *post hoc* lurks in any causal interpretations we might seek to impose on statistical data. Statistical data often suggest causal relationships. But be careful: do not leap to such suggested conclusions too quickly.

Examples for Discussion

1. An illustration of the difficulty of imposing causal interpretations on statistical data is found in the race-IQ controversy.[17] One thesis says that the reliably obtained racial differences in IQ are primarily genetically determined. That is, genetic differences are seen as the primary *cause* of measured IQ differences. The polar-opposite viewpoint is based on exactly the same evidence. This viewpoint attributes racial-IQ differences to environmental differences between the races. Because of the *post hoc* nature of the evidence and the explosive potential of the conclusions, some theorists have concluded that it is impossible, at the present time, to determine the cause(s) of racial-IQ differences. But they also conclude that perhaps the question should not even be investigated.

2. Another example for discussion is the theory of Dr. Melvin Page, a dentist in St. Petersburg, Florida, and head of the Biochemical Research Foundation. "As far as I know, he [man] and a certain species of ant are the only ones who use an animal secretion after the age of weaning."[18] Page thinks milk is fine for babies before weaning but that after

weaning, it is a dangerous cause of colds, sinusitis, colitis and cancer. The doctor points out that more people die of cancer per capita in Wisconsin than in any other state – and Wisconsin produces more milk than any other state. Dr. Page fears that unless we stop drinking this animal secretion and reform our diets in other curious ways, the Anglo-Saxon race will continue to degenerate faster than certain "primitive" races (which he does not specify).[19]

3. Dr. Manfred Sakel announced, in 1927, that schizophrenia could be treated by insulin doses. The insulin, in turn, produced convulsive shocks. Subsequently, numerous physicians adopted the practice of treating schizophrenia by means of administering electric shock treatments (without insulin). Some time later, Dr. Sakel regretfully explained that it was the insulin treatment that was beneficial in the treatment of schizophrenia because of its effect on hormone levels, and that electric shock treatment could be harmful.

Notes

1 The example is drawn from Brian Skyrms, *Choice and Chance: An Introduction to Inductive Logic* (Belmont, California: Dickenson, 1966), page 9. This fine little book is a very good introduction to the fascinating world of inductive logic and probability theory.

2 Stephen K. Campbell, *Flaws and Fallacies in Statistical Thinking* (Englewood Cliffs: Prentice-Hall, 1974), Chapter 11.

3 Clear and concise elementary introductions to the rules of the probability calculus are, in order of increasing sophistication:

    a Brian Skyrms, *Choice and Chance*, Chapter IV (See note 1 above).

    b Henry E. Kyburg, Jr., *Probability and Inductive Logic* (London: Collier-MacMillan, 1970), Chapter 2.

    c Arthur Smullyan, *Fundamentals of Logic* (Englewood Cliffs: Prentice-Hall, 1962), Chapter 5.

4 Antoine Arnauld, *The Art of Thinking*, translated by James Dickoff and Patricia James (Indianapolis: Bobbs-Merrill, 1964), III, 20.

5 Amos Tversky and Daniel Kahneman, "Belief in the Law of Small Numbers," *Psychological Bulletin* 76 (1971): 105–110.

6 Tversky and Kahneman, page 106.

7 Daniel Seligman, "We're Drowning in Phony Statistics," *Fortune*, November (1961).

8 Campbell, page 56.

9 In addition to Campbell, the reader may wish to consult Darrell Huff, *How to Lie with Statistics* (New York: Norton, 1954).

10 Leonard Silk, *Economics in Plain English* (New York: Simon and Schuster, 1978), page 190.

11 David Hackett Fisher, *Historians' Fallacies* (New York: Harper & Row, 1970), page 166.

12 Harold L. Larabee, *Reliable Knowledge* (Boston: Houghton Mifflin, 1954), page 368.

13   Huff, page 98.

14   Hans Zeisel, *Say It with Figures*, 5th ed. (New York: Harper & Row, 1968), Chapter 9. See also H.A. Simon, "Spurious Correlation: A Causal Interpretation," *Journal of the American Statistical Association* XLIX (1954): 467–492.

15   Fisher, page 177.

16   For further information about the model of argument involved in *post hoc*, student and instructor alike are urged to turn to philosophical literature on causation. A standard text is *Causation and Conditionals*, edited by Ernest Sosa (London: Oxford University Press, 1975).

17   This controversy was recently brought into the public eye through publication of the work of A.R. Jensen, "How Much Can We Boost IQ and Scholastic Achievement?" *Harvard Educational Review* 39 (1969): 1–123.

18   Melvin Page, *Degeneration — Regeneration* (St. Petersburgh, Florida: Bio-Chemical Research Foundation, 1949).

19   Martin Gardner, *Fads and Fallacies in the Name of Science* (New York: Dover Publications, 1965), pages 222–223.

Chapter Five
The *Ad Verecundiam*

We mentioned the *ad verecundiam*, or appeal to authority, in Chapter 2 (page 25). The danger of this fallacy is ever-threatening in argument. The bold rejoinder, "Authority X says so!" is an argument clincher; too often, it is devastatingly effective as a decisive move to counter or divert an opponent's thrust. Any move in argument towards the citation of authorities is very hard to know how to deal with. The reasoned evaluation of arguments from authority has been largely neglected by logicians. Indeed, numerous logic texts indicate that *any* appeal to authority is an *ad verecundiam*, a fallacious move in argument.

But the idea that any appeal to authority is fallacious is itself a substantive thesis about the *ad verecundiam*. We think the thesis is wrong because, as we will try to show in this chapter, *some* appeals to authority are legitimate moves in argument. Sometimes an authority can be a genuine source of knowledge; genuine knowledge may be legitimately referred to. Our thesis is that appeals to authority are sometimes right and sometimes wrong; thus it behooves us to show, exactly, the conditions under which such an appeal can go wrong. Let us look to this question first. We will then take up the related question, how an appeal to authority can be a right argument. We will see that a correct *ad verecundiam* argument represents a third type of argument, neither inductive nor deductive in nature. We will call this type *plausible argument*.

1. Five Forms of *Ad Verecundiam*

Albert Einstein got into a lot of trouble because, being a sincere man, he would try to give an honest answer to questions of a political, ethical or religious nature from reporters or other informants. Einstein was not an expert in politics, ethics, or religion but, because of his high prestige as an authority in physics, people tended to take his pronouncements in

86

these areas very seriously, and his remarks often made headlines. So it is with experts. We value their opinions in area A because of their expertise in area A. But by a kind of "halo effect", their pronouncements in areas B or C, which may be altogether unconnected with area A, are escalated in credibility. Appealing to expertise is not wrong – but such an appeal can go wrong in numerous ways if we are not very careful. The problem is aggravated by the narrowness of many fields of expertise.

Even the most impeccable appeal to expertise is limited, for experts can be, and are, wrong. So we should have a certain mistrust of taking an expert at his or her word without troubling to examine other evidence. For example: in the middle ages, the word of the philosopher Aristotle was very often taken as virtually sacrosanct; his was the last word on many subjects. Subsequent developments in experimental science tended to erode Aristotle's authority. Perhaps there is some truth in the observation that, ever since the erosion of Aristotle's authority, Western society has tended to be highly suspicious of authorities. And the misuse of authority by vested interest groups in society is a nuisance that is often remarked on by historians and political commentators.

But in the modern world there are many narrow, highly specialized fields of knowledge. Often, through lack of time or resources for independent investigation by everyone involved, we are forced to accept the sayso of suitably qualified experts. You may question your physician's diagnosis, or ask for a second or third opinion – but in the end, you listen to your doctor. It is rational to trust the diagnosis of well-qualified practitioners, which seems to have more authoritative value than your own inexpert speculations. It may be hard for us to understand subjects like relativistic physics or the economics of inflation. So if the body of experts tends to agree on certain propositions in those subjects, it is reasonable to think that, for all we know, those experts are likely to be right.

An appeal to authority is rarely – if ever – conclusive in every way. But it does alter the organization of the preponderance of evidence, the burden of proof. If we have to make a significant diplomatic or political decision regarding nation X, it is wise to take into account the testimony of someone who knows the relevant facts about the culture and current political affairs of X. It makes better sense to be guided by evaluated information from knowledgeable sources than to make a decision on the basis of ignorance or uninformed guesswork. However, some experts are more reliable than others, and some would-be experts are no experts at all. Many an appeal to expertise can mislead.

What can go wrong in appealing to authorities? What logic textbooks call *ad verecundiam* is, in reality, an umbrella term for a number of specific errors of argument. Below we cite five conditions that must be met for an appeal to authority to avoid fallacy. Each condition implies one particular way in which such an appeal can go wrong.

Condition I The authority must be interpreted correctly.

In an appeal to authority, common pitfalls are misquotation, mis-documentation and inadequate rendering of the expert's expressed judgment. Subtle changes in emphasis can be very important. If the claim is in the form of a direct quotation, it must be quoted in proper context. De Morgan warns us of six problems in the citing of sources.[1]

1. It is not uncommon, in disputation, to fall into the fallacy of making out conclusions for others by supplying missing premises. One says that A is B; another will take for granted that he or she must believe B is C, and will therefore consider the expert as maintaining that A is C.
2. As to subjects in which men go in parties, it is not very uncommon to take one premise from some individuals of a party, another from others, and to fix the logical conclusions of the two upon the whole party: when perhaps the conclusion is denied by all, some of whom deny the first premise by affirming the second, while the rest deny the second by affirming the first.
3. Quotation is [not] obligatory, though highly desirable: but the reader must remember, when there is only citation, that it is not the author cited who speaks, but the person who brings [the author] forward. . . . If the citer be honest, the passage in question exists: if judicious, it is to the effect stated.
4. Perhaps the greatest and most dangerous vice of the day, in the matter of reference, is the practice of citing citations, and quoting quotations, as if they came from the original sources, instead of being only copies.
5. Unjustifiable as unnoted omissions may be, still more so are additions and alterations.
6. Omissions of context, preceding or following the quotation, may alter its character entirely: and this is one of the most frequent of the fallacies of reference, both intentional and unintentional.

There is also the danger that an outdated expert opinion may be cited. If the expert has changed his or her mind, the earlier opinion loses much of its force. In other cases, the appeal to authority may be ruinously misleading if certain qualifications added by the expert are omitted or suppressed. These qualifications are particularly important where the expert's probability-assessment is a conditional one; an expert might make an argument conditional *on certain contingencies*. If these contingencies are not preserved in the report of the expert's judgment, the report is misleading. If the authority delivered a judgment in technical terminology, that judgment might have to be rendered into a more readily intelligible form. This can be especially important where a panel of experts is attempting to make an interdisciplinary prediction. Injudicious technical jargon can be disruptive and should be avoided. Yet many of the strikingly fallacious forms of the appeal to authority trade on ambiguities, vagueness or misunderstanding in the translation of technical terminology to a usable language.

Condition II The authority must actually have special competence in an area and not simply glamour, prestige or popularity.

Many of the appeals to authority typical of treatments in logic textbooks are really appeals to prestige or popularity. It is difficult to decide when a putative expert actually has the required special competence; this difficulty cannot be overemphasized. In many areas, it may be doubtful that there are any genuine authorities. In other areas, the means of evaluating the degree of expertise may be meagre or only dimly understood. Even in those provinces where criteria of special competence is relatively exact, there could yet be difficulty in rating the degree of expert competence. Some indices of competence are suggested here. Obviously they are not standardized; nor is it very likely that an exact set of criteria will be forthcoming in the near future. For all its roughness, a provisional rating might be based on three criteria.

  i) Previous record of predictions.
 ii) Where a previous record of predictions is not available, we may have a record of hypothetical predictions. This might be some sort of test that the purported expert may have undergone; the test could indicate a rating of the expert's degree of competence.
iii) Access to a record of other sorts of qualifications – degrees, professional qualifications, testimony of colleagues or other experts – or a record of familiarity with the area of expertise. The criteria, it is expected, will vary considerably from area to area.

We can ask a number of specific questions to evaluate our three criteria.

First: if holding an important post or position is relevant to expertise in the particular area in question, we can ask whether the post held is a desirable one, and how the appointment to the post was made. Was the expert appointed by a committee of experts? Or was the appointment a matter of politics or popularity? Second: if a degree is relevant, what school is the degree from? Is it an earned degree or an honorary one? What are the parameters of judgment of a "good" school? Third: what sort of awards or grants has this person won to sponsor his or her research? Fourth: has the expert published findings in books or journals? If in journals, have these publications been refereed by other qualified experts? If the expert has published books, have they been favourably reviewed by other experts? By experts who might be expected to be somewhat opposed to the published views? Many specialized book and article review indexes can be consulted to obtain critical and informed comments and evaluations. Do other experts in the area refer to or acknowledge our expert's findings? Fifth: what do the expert's professional colleagues think about him or her? Are the colleagues consulted generally thought to be honest and reliable?

Condition III The judgment of the authority must actually be within the special field of competence.

The most blatant of the fallacious appeals to authority occur when a legitimate expert in a given area makes a judgment that does not fall within his or her field. Added credibility is given to his or her judgment simply because he or she is an expert in some area. In many cases, particularly interdisciplinary cases, it may be difficult to decide whether your question actually falls within a domain that admits some particular kind of expertise. There are areas in which the judgment of an expert is considered more valuable than the judgment of a lay-person, but many of these areas do not have well-defined criteria of expertise.

An economic anecdote will serve as an example. The famous economist John Maynard Keynes "was having dinner with Professor Max Planck, the mathematical genius who was responsible for the development of quantum mechanics, one of the more bewildering achievements of the human mind. Planck turned to Keynes and told him that he had once considered going into economics himself. But he had decided against it – it was too hard."[2]

Condition IV Direct evidence must be available in principle.

For an appeal to authority to be adequate, the authority must base his or her judgment on actual, relevant and objective evidence within the area concerned. The evidence need not be fully available; as we have said, it may not even be exactly specifiable. But where there is doubt, the authority must give some evidence that his or her judgment is objective.

For example, suppose we have a panel of experts in a given domain of experience; suppose also that the judgment of one panel member on a certain question falls well outside the range of consensus. In an evaluation of that particular judgment, evidence must be made available for the evaluation of the other experts on the panel. Similarly, where we have reason to believe that our one panel member may not be trustworthy, we should call for an evaluation of the degree to which his or her judgment is based on relevant evidence.

Condition V A consensus technique is required for adjudicating disagreements among equally qualified authorities.

An obvious problem with expert consultation is that authorities are notorious for their ability to disagree. Yet since two (or more) heads are better than one, the rational method, when faced with disagreement, of obtaining the required information (inconclusive though it may be) is to take a consensus. In fact, a second consensus, after mutual inspection of the grounds of disagreement, might even have a convergence effect. The usefulness of establishing a method for the process of consensus among experts has become increasingly apparent recently; considerable effort has been spent to develop techniques for this purpose. Most notable have been the efforts of Norman Dalkey and others at the Rand Corporation to develop the so-called Delphi technique. The Delphi technique is a method for the systematic solicitation and collation of informed judgments on a topic. The Delphi procedure consists of a set of sequential questionnaires, interspersed with summarized information and opinion feedback derived from earlier responses. The technique is intended to prevent professional status and high position from forcing judgments in certain directions; such politicking frequently occurs when panels of experts meet. The intention is to ensure that changes in estimates reflect national judgment; too often, those changes were reflections of the influence of opinion leaders and trend-setters.

In the Delphi procedure, each member of a panel of experts in an area is asked to write his or her individual responses to questions; those responses are made available to all respondents. Any responses that have fallen outside a certain range (often the interquartile range, the interval containing the middle fifty per cent of responses) elicit requests for justification or evidence. Next, a second round of responses is collated; it is hoped that this second round will cause convergence of opinions – that is, it will reduce the spread of opinion. This technique is regarded as an adjunct to face-to-face panel discussion.

Though not illegitimate, opinion-based arguments, like the appeal to authority, are no substitute for a direct appeal to available evidence. Reliance upon expertise is not a substitute for direct evidence. But it seems to be an indispensable adjunct to the reasonable and enlightened use of direct evidence. It is plausible to conjecture that, as direct access to data becomes available, the appeal to authority will diminish in respectability.

A Case Study: The Shroud of Turin

The Shroud of Turin is a fourteen-foot-long piece of cloth, apparently marked with blood stains. The stains outline the remarkable image of a five-foot-seven-inch man with braided hair and a beard. The Shroud of Turin is believed by some to be the burial shroud of Jesus Christ. It first surfaced in France in the fourteenth century; it is now kept in a silver box in a church in Turin, Italy.

Nine items of information about the status of scientific studies of the shroud were reported in a recent newspaper article.[3]

   1.   Scientists examining the cloth have found no trace of blood or pigment

indicating that the enigmatic figure was painted on the woven linen cloth. Joseph Accetta, one of thirty-four researchers permitted to examine the shroud in 1978, says no one is yet able to explain the figure on the material. Scientists have found details so fine that they are invisible under the conventional microscope.

"We still haven't identified the image on the cloth," Accetta told reporters while in Ottawa recently to talk about the shroud.

Accetta, a laser physicist with Lougheed Aircraft in Albequerque, New Mexico, says some suggest a radiant burst of energy somehow scorched the cloth. But as one observer asked, "How could a body, cold in its tomb, produce that kind of precisely detailed burn?"

Accetta says there is a contradiction in data and notes that the imprint affects only the topmost surface of the linen fibres.

2. Accetta, like his colleagues, has donated his spare time and specialist knowledge to solving one puzzle.

"For the 1978 expedition we were interested in identifying the chemistry of the image," he said. "We wanted to find out what makes the image different from the surrounding cloth."

Their experiments included elemental analysis by X-ray fluorescence, ultraviolet and infra-red ray analysis to search out possible organic compounds. As well, they exposed the shroud to ion microprobe – a kind of super electron microscope that can detect tiny particles and prove positively the absence of paint.

3. Research in the past ten years has shown that the cloth fibres and weave are typical of those manufactured around the time of Christ. A Swiss criminologist also found pollen grains on the shroud coming from plants that exist almost exclusively near the Dead Sea region of the Holy Land.

4. The Roman Catholic Church has made no official claims about the authenticity of the cloth, although it vigorously denounced the shroud as a forgery in the fourteenth century.

5. Skeptics call the material a cunning fake owing to its timely appearance in history – 1353 in France. In the fourteenth century, Christian imagination flourished as "relics" appeared in abundance: a feather from St. Michael's wing, a ray from the star of Bethlehem and more. Religion had become a money-making racket.

6. Experts say that, with the new technique, they need only some threads and two small samples removed earlier during repairs in 1973. However, the crucial Carbon 14 dating test, which would determine how old the cloth is, has not yet been approved.

7. The shroud-man's back is covered with dumbell-shaped marks where the skin apparently was broken by flagellation; the marks correspond with the leaded Roman whip. Also multiple puncture wounds suggest that a clump of thorny twigs – and not the circle of thorns depicted by most artists – was pressed down on the head. Nor had the shroud-man suffered the traditional breaking of the leg bones by a mallet.

8. Most puzzling, however, is that crucifixion nail wounds were found on the image's wrist and not the palms, as reported in the Gospel.

Science has proven this to be anatomically correct. Experiments performed by a French doctor in the 1930s with corpses and amputated limbs showed that the weight of a human body would have caused nails driven through the palms to rip the flesh up between the fingers, since no bones would bar their way. Wrist-nailing would have ensured that the body stayed on the cross.

9. More amazing still is a medical detail which sindonologists believe diminishes the likelihood of a fake.[4] If a nail pierces the wrist between two specific wrist bones, it touches a nerve that automatically causes the thumb to flex across the palm, leaving it unseen to anyone looking at the back of the hand. Neither thumbs can be seen on the hands of the shroud-figure.

At the conclusion of the article, scientists acknowledge that they will never be able to prove beyond a shadow of doubt that the figure on the shroud is that of Christ. Accetta is quoted as saying, "I would be satisfied if at the time [1978] we had got some reasonable identification of the surface chemistry [of the cloth]. That's about as far as we can reasonably go."

Much has now been written about the Turin Shroud. The report outlined above illustrates the fascination of the subject – and the inherent difficulties in assessing the evaluations of the scientific experts from the diverse technical fields that bear on questions of the likely genuineness of the Shroud.

We suggest two interesting projects. First: using only the information we have given, evaluate the quality of a case concluding that the Shroud is genuine. Use the five conditions for the adequacy of an appeal to authority, which we have outlined. What are the strongest and weakest points of such an argument? Warning: try to state as precisely as possible the conclusion(s) you propose to evaluate. Second: obtain more information concerning appeals to scientific expertise in relation to studies of the Turin Shroud. Use this material to study the implementation of our guidelines for an appeal to authority.

Exercises

Identify any specific errors or shortcomings in the following appeals.

1. Babs Bazersky, noted author of *Molasses: Key to Bodily Energy* and expert on nutrition, recommends eating three pounds of molasses every day. Thousands have discovered new energy and release from acne and other disorders by this regimen. You should try it!

2. The noted art historian, Trevor Baloon, thinks that "Frog Recumbent" is the finest statue in Western Art. This seems odd, but Baloon is an expert.

3. Dr. K. Margaret Baxter-Guano, the eminent obstetrician, has stated in a press release that wantonly doing away with the unborn by allowing abortion on demand violates the principle of the sacredness of human life. She is the author of four books

on obstetrics and is a Fellow of the Royal College of Physicians and Surgeons. Hence we can say with authority that abortion must be wrong.

4. The famous philosopher Socrates was noted for his injunction "Know thyself." Here we have an early exponent of psychiatric examination and therapy.

5. Professor Bagdolio, medical doctor and expert on the Shroud of Turin, claims that the representation on the shroud is not a painting or a humanly designed artifact. Rather, he believes that it must be the natural imprint of a human body. The noted historian Professor Lenzgrinder, who is also a well-known authority on the shroud, would have us believe that if the representation on the shroud is the imprint of a human body, it is that of Jesus Christ. We may conclude that the shroud represents a picture of Christ.

A Second Case Study

Professors Robert Gordon Shepherd and Erich Goode conducted a study to find out whether the scientists whose work is cited by the press are those who have conducted original research on a given subject.[5] They chose to examine scientists who were conducting research on marijuana use. There have been public and scientific controversies on this subject; scientists are debating whether marijuana causes brain damage.

Shepherd and Goode based their study on a sample of 271 scientific articles and 275 newspaper and magazine articles. The articles were published between 1967 and 1972. The scientific articles were selected from *Index Medicus*; the second set of articles was chosen from a range of well-known American magazines and newspapers. To gauge the standing of a scientist and his or her work in the scientific community, they calculated the number of citations that scientist had received in the *Science Citation Index*. They studied two types of popular articles on marijuana, both of which involved scientific experts. First there was the "news story", which gave results of a scientific study; second, the "summary article" summarized the state of research on some marijuana-related issue.

Their findings were interesting. They found that there were very few "news story" articles and even fewer scientific publications mentioned in the news stories, but that those publications cited did represent well-known research in the scientific community. It would appear that even topical scientific research receives very little attention from the press, but what coverage it gets is a first-rate selection in terms of its frequency of citation by other scientists.

But most of the popular articles were of the summary type. The correlation of citations (those in the press with those in scientific journals) formed a dismal picture. Of the ten marijuana researchers most often cited in the scientific literature, for example, only one appeared in the list of the ten most publicized authorities in the press. Of the ten press "authorities" seven had not published anything at all in the scientific literature. The results overall indicated that the press was not interested in seeking out the views of the most scientifically influential researchers. Spokesmen cited in the press seem to have been selected because they were the administrative head of some health-related agency, faculty, or institute.

It appears that the administrator or director is thought, by the public, to be the "boss" of the organization; his title is most effective as a citation that carries with it a weighty endorsement of authority. This would also seem to be the reporter's point of view. Why quote a mere working scientist when you can quote the head of a whole institute?

According to Shepherd and Goode, there is another factor that contributes to the press's tendency to rely on "authorities" who are not really authorities in a specific area of research. Shepherd and Goode theorize that it is not easy for most non-experts to distinguish between the specialized scientific disciplines involved. Being a doctor, for example, seems to make any physician an "instant expert" on all aspects of drug use.

The irony of it, Shepherd and Goode conclude, is that reporters generally pride themselves on going directly to the primary sources when they write stories. In other areas, like athletics or the state of the novel, they will interview the players and the novelists themselves. But to report scientific research involves much work and study; the reporter must go to the library and search out technical reports, summaries, indexes, and so forth. It saves time – and possibly fruitless effort – to simply talk to someone who seems conspicuous, someone with credentials, someone the public will readily acknowledge as an expert.

The study indicates that there are probably very widespread lapses into the fallacious *ad verecundiam*, where issues of public controversy and importance are influenced by appeals to scientific expertise in a specific area of research.

Exercise

Find a magazine or newspaper "summary report" of research related to an issue of public controversy. Evaluate some of the weaker points in the part of the argument where authorities are appealed to. Make use of factors cited in Conditions I to V and the Shepherd-Goode report in your analysis.

2. Towards the Model of Argument Involved

We have looked at some failings of the *ad verecundiam*. We turn now to the question of what kind of argument could constitute a correct argument from authority. We looked at deductive and inductive arguments in preceding chapters, and it is natural to think that one of these types of argument is at issue. Indeed, both of these suggestions have been made. C.L. Hamblin suggests that we start from the deductively correct form of argument, "Everything X says is true, and X said that *p*, therefore *p*".[6] But, Hamblin says, we can still expect to find weaker, though not quite fallacious, forms of the argument from authority. Wesley Salmon disagrees with Hamblin. Salmon asserts that the appeal to authority could not be deductively valid, because the premises of the above argument could be true while the conclusion is false. Salmon's reason: that no authority, at least outside considerations of theology, is infallible or omniscient.[7] Although we will see that deductive logic can be partially utilized in evaluating appeals to authority, it seems clear that many reasonably good appeals to authority are not deductively valid arguments. According to Salmon, the appeal to authority is better thought of as belonging to an inductive argument; the appeal can be inductively correct. He suggests the form: "The vast majority of statements made by X concerning subject S are true; *p* is a statement made by X concerning subject S; therefore *p* is true." Let us examine the suggestions of Hamblin and Salmon.

Remember that most useful appeals to expertise are, by their nature, highly variable and fallible. We have seen the various ways they can quickly become unmanageable. Such appeals depend on what sort of authority is consulted, and on how the authority's opinion squares with other authorities. Perhaps everything a fallible but expert authority, X, says is true; X says that *p*. But it need not always follow deductively that *p* is true. Why not? First, consider this: perhaps everything another fallible but equally reliable authority, Y, says is true; Y says that not-*p*. If Y's pronouncement is treated as deductively correct, it follows that not-*p* is true. But when we apply the same assumption of deductive correctness to X's statement, it follows from what X said that *p* is true. Given both presumed inferences, we have it that *p* is both true and not-true. But this is a flagrant logical contradiction! So we must conclude that many a reasonably good argument from authority is not deductively correct. Because authorities maintain conflicting pronouncements, it hardly follows that what they say is both true and false.

Perhaps, if we find an authority who is perfectly reliable in every possible instance – an Oracle or Prophet – the argument from authority can be treated as a kind of deductive argument. But most of the authorities we appeal to – in science, medicine, and so forth – are not infallible. Thus deductive logic does not provide the proper forum for the argument from authority.

What about inductive logic? As an inductively correct argument, an appeal to authority might look like this: "The vast majority (more than 90%) of statements made by X concerning subject S are true; *p* is a statement made by X concerning subject S; therefore Pr (*p*) > $^9/_{10}$." But we have a problem – a familiar one. For suppose we have another expert who claims that the probability of not-p is very high, say $9/10$. This gives us another inductively correct argument: "The vast majority of statements made by Y concerning subject S are true; $\neg p$ is a statement made by Y concerning subject S; therefore Pr $(\neg p) > 9/10$." Remember the negation rule of inductive logic: Pr $(\neg p) = 1 - \text{Pr}(p)$. We apply this rule to the pronouncements of our authorities: if Pr $(\neg p) > 9/10$, then Pr (p) $\leqslant 1/10$. So here again we have a contradiction. For it cannot be true that the probability of p is both greater than $9/10$ and less than or equal to $1/10$.

Thus we must conclude that not every reasonably good argument from authority is inductively correct. Legitimate authorities do disagree, but it hardly follows that their pronouncements can have both a high and a low probability value at the same time. The laws of inductive logic do not allow it, any more than the laws of deductive logic allow contradictions to be true.

The argument from authority may be inductively correct if our authorities are so reliable that, as a group, they are always *perfectly* regular in their assessments of probabilities. But to think we will find these perfect authorities is somewhat idealistic.

Unlike deductive and inductive argument, which are based exclusively on propositions, the *ad verecundiam* is a *subject-based* form of argument. That is, the argument from authority X on proposition p may be assigned an entirely different value than the argument from authority Y on the very same proposition, p. More than the truth-value or the probability-value of the proposition itself is at stake. We must consider *which* authority backs p. What one authority asserts may be quite different from or even contradictory to what is asserted by another authority.

We must also distinguish between the *de facto* and the *de jure* appeal to authority. The *de facto* appeal is the appeal to "factual" testimony or knowledge; it is characteristic of the usual appeal to expertise. The *de jure* appeal is very different. Sometimes we refer to the titular, legislative, or structure-of-command authority invested in an individual; this authority is not conferred because of the individual's particular expertise. For example, ministers and ships' captains are invested with the authority to perform the marriage ceremony. In the proper circumstances, if such a duly constituted authority says, "You are now man and wife" then indeed you *are* man and wife; saying it makes it so. Notice that in such a situaton there is at least something of the logical tightness that characterizes deductive arguments. But such *de jure* pronouncements by authority are quite different from the *de facto* arguments from the authority of expertise in a subject-matter that we have been examining.

Many a fallacy of *ad verecundiam* occurs in the confusion between *de jure* and *de facto* authorities. Over the years, many official sources have been by custom invested with infallibility and finality. Religious authority, especially, has been used in this fashion to suppress heretical, disloyal, or radical opinions. But in a *legitimate de facto* appeal to authority, criteria of special competence and expertise must be met. And, as we saw in Condition IV, hard evidence must be available, and the putative expert must be able to cite it, if necessary.

We have now eliminated deductive and inductive logic as models of argument appropriate to studying the *de facto* argument *ad verecundiam*. Where should we look next? We will have to find a third type of argument. But first, let us examine the problem more closely.

3. Inconsistency

We saw that deductive logic does not provide much help when we are confronted with false premisses. And deductive logic is even less helpful when the premisses are inconsistent. This is because an inconsistent set of premisses can entail any conclusion you like. We defined deductive entailment as follows: a set of premisses deductively entails a conclusion where it is logically impossible for the premisses to be true and the conclusion false. But let us suppose that our premisses are inconsistent – that is, they are logically impossible. It will always be logically impossible for the premisses to be true and the conclusion false. Reason: if a set of propositions is inconsistent, then our premiss can never be true. Adding another proposition to the premiss won't help: the set will still be inconsistent. For example, take the set of propositions {John Woods has red hair, John Woods does not have red hair}. This pair of propositions is inconsistent: it is not logically possible for both members to be true. Adding more propositions to this inconsistent set is not going to *remove* the inconsistency. The only way to remove the inconsistency is to take one of the propositions away. When we are confronted with inconsistency, deductive logic is not much help. It can *identify* the inconsistency, but it does not tell us how to remove it or what to do next.

Nor is inductive logic any help. We remember that the probability of p given q (conditional probability) is defined as the probability of $(p \wedge q)$ divided by the probability of q. But if the given q is inconsistent, its probability value is zero. And dividing by zero is simply not allowed in mathematics. So, again, we are stuck.

But we need to know how to argue when confronted by inconsistency. As we have seen, the authorities we appeal to often disagree; what they say is therefore collectively inconsistent. Moreover, in the case of the circumstantial *ad hominem*, we are confronted by an allegation of inconsistency; we have to know how to deal with such an accusation. Therefore we need a third approach to argument: one that is neither deductive nor inductive, and one that will help us with *ad verecundiam* and other fallacies that involve dealing with sets of premisses that may be inconsistent.

Before we look for a new approach to argument, we should make sure that we can use the tools at hand to identify inconsistencies. Deductive logic is the most useful tool. Basically, an inconsistent set of propositions is one where it is not possible for all of the propositions to be true at once. The most elementary form of inconsistency is that represented by $p \wedge \daleth p$. It is the very nature of classical negation that p and $\daleth p$ cannot both be true.

p	$\daleth p$	$p \wedge \daleth p$
T	F	F
F	T	F

Generally, the nature of inconsistent sets of propositions is just this: if you form a conjunction of those propositions, the resultant proposition cannot be true. In a truth-functionally inconsistent set, its conjunction will be false in every possible row of the truth-table. Take the set $\{p \vee q,$ $\daleth p, \daleth q\}$. Is it consistent or inconsistent? To decide, we draw up a truth-table for the conjunction formed from all three propositions.

p	q	$p \vee q$	$\daleth p$	$\daleth q$	$(p \vee q) \wedge$ $\daleth p \wedge \daleth q$
T	T	T	F	F	F
T	F	T	F	T	F
F	T	T	T	F	F
F	F	F	T	T	F

For a conjunction to be true, each conjunct must be true. But if we scan the rows for our three conjuncts – $p \vee q$, $\daleth p$, and $\daleth q$ – we see that in no row are all three all true. Hence $(p \vee q) \wedge \daleth p \wedge \daleth q$ is false in every row. We conclude that it is inconsistent.

Generally, we can always determine whether a set of propositions is truth-functionally inconsistent by simply drawing up a truth-table for the conjunction of its members. If the final column is false for each conjunct, the set is inconsistent. Otherwise, it is consistent.

Exercise

Determine whether each of the following sets of propositions is inconsistent:

(a) $\daleth p, \daleth q, p \supset q$;

(b) $p \supset q, q \supset r, \daleth (p \vee q)$;

(c) $p \supset q, r \supset q, \daleth q, r \wedge p$;

(d) $p \vee q, \daleth (p \wedge q)$;

(e) $p \equiv (q \vee r), \daleth p, \daleth q, r$;

(f) $p, q, r, p \supset q, q \supset r, p \supset s, r \supset \daleth s, p \wedge \daleth s$

Now that we can identify inconsistency, we need a way to deal with it. As we saw above, adding propositions to the inconsistent set is no help. We need a way to *remove* some of the propositions from the set. We introduce the notion of a *maximal consistent subset* of an inconsistent set of propositions – that is, a subset that is consistent but that immediately becomes inconsistent if any more propositions from the original set are added. Take the set $\{p \lor q, \lnot p, \lnot q\}$, which we know is inconsistent. $p \lor q$ is a consistent subset, but it is not a maximal consistent subset, because we can add $\lnot p$ to it and the result will still not be inconsistent. To check, do a truth-table for the conjunction $(p \lor q) \land \lnot p$, and you will see it is consistent. But can we go any further? Suppose we add $\lnot q$. This gives us the original set, which was inconsistent. So that is as far as we can go. Therefore $\{p \lor q, \lnot p\}$ is a maximal consistent subset. By similar reasoning, we can see that $\{p \lor q, \lnot q\}$ is a maximal consistent subset of the original set, and so is $\{\lnot p, \lnot q\}$. And these are the only ones.

The test for a maximal consistent subset of some original set of propositions is quite simple. Take each element of the original set one at a time; test each for inconsistency, then keep adding other members from the original set, one at a time, until you reach an inconsistent set. What you had *before* you attained an inconsistent set was a maximal consistent subset. A maximal consistent subset is one that is as big as it can possibly be without becoming inconsistent.

There is an easy way to find all the maximal consistent subsets in the truth-table of a set of propositions. Scan each row of the truth-table and list each proposition that is true in that row; this will give you all the *consistent* sets. But ones that are included in other sets may be eliminated as candidates for maximal consistent sets. Consider the truth-table for our example, $\{p \lor q, \lnot p, \lnot q\}$.

	p	q	$p \lor q$	$\lnot p$	$\lnot q$
(1)	T	T	T	F	F
(2)	T	F	T	F	T
(3)	F	T	T	T	F
(4)	F	F	F	T	T

Scanning row (4), we see that $\{\lnot p, \lnot q\}$ is true. Then, in row (3), we list $\{p \lor q, \lnot p\}$ as true. Next, in row (2) we list $\{p \lor q, \lnot q\}$. Finally, we list $\{p \lor q\}$ in row (1). But notice that it is already included in rows (2) and (3) as well; we do not need to consider it. It is not a *maximal* consistent subset. By looking at the truth-table, we can be assured that we have listed all the maximal consistent subsets for any given set of propositions. In our example, there are only three.

Exercise

Find all the maximal consistent subsets of each of the three sets of propositions you found to be inconsistent in the exercise on page 99.

4. Plausibility Screening

Suppose that an authority or group of authorities has made a number of pronouncements; suppose further that we can rate the plausibility of these authorities in some comparative way. Suppose also that the set of propositions vouched for by the authorities is collectively inconsistent. Is there a rational way to deal with the obstacle of an inconsistent set of givens? One procedure has been developed by Nicholas Rescher; it is called *plausibility screening*.[8] Rescher's method is to scan the maximal consistent subsets and give preference to those that include the maximum number of high-plausibility elements. Or the process can work in other ways. For example, for some purposes of evaluation, we might give preference to the sets that include as few low-plausibility elements as possible. It depends on whether our goal is to maximize overall plausibility or minimize overall implausibility. *The two policies are not identical.*

Suppose that we are confronted by a fragment of what appears to be a thirteenth-century manuscript on logic. It has been examined by three experts on historical manuscripts on the logic of this period, Professors X, Y, and Z. Let's say that we can rate their respective pronouncements on a scale of one to ten as follows: X has a comparative plausibility value of eight (highly reliable); Y has a value of five (fairly reliable); and Z has a value of two (somewhat reliable). Professor Y ventures the pronouncement that the manuscript was authored by William of Sherwood, the thirteenth-century logician, or by Roland of Poland, a more obscure figure, about whom little is known. Professor X asserts that if the document was authored by William of Sherwood then it would definitely make reference to Aristotle's doctrine on logic. But, he adds, it does not anywhere make reference to Aristotle's doctrine. Professor Z points out that if the document was authored by Roland of Poland, then from what we know of Roland, it would include references to Aristotle's doctrines, too.

We are supposing, then, that each of these authorities vouches for the following propositions:

$$X \text{ (value of 8):} \quad p \supset r, \urcorner r$$
$$Y \text{ (value of 5):} \quad p \lor q$$
$$Z \text{ (value of 2):} \quad q \supset r$$

The set of given propositions, $\{p \supset r, \lnot r, p \lor q, q \supset r\}$, is inconsistent, as a truth-table will show. But the set has four maximal consistent subsets:

(1)	$\{p \lor q, p \supset r, q \supset r\}$	Reject $\lnot r$.
(2)	$\{p \lor q, p \supset r, \lnot r\}$	Reject $q \supset r$.
(3)	$\{p \lor q, q \supset r, \lnot r\}$	Reject $p \supset r$.
(4)	$\{p \supset r, q \supset r, \lnot r\}$	Reject $p \lor q$.

Notice that (1) and (3) both reject the highly plausible pronouncements of X. Therefore we can eliminate both (1) and (3) as preferable subsets. Looking at the remaining two subsets, we see that we have a choice between rejecting $q \supset r$ (plausibility 2) and $p \lor q$ (plausibility 5). Our policy, as we mentioned, is to maximize plausibility; therefore, we want to reject any alternative that excludes propositions of relatively high plausibility. So the choice here is straightforward. We reject (4) because it excludes $p \lor q$, a proposition that is more plausible than $q \supset r$, the proposition excluded by (2). All told, then, the most plausible maximal consistent subset is (2). The rational way to react to inconsistency in this instance is to accept the pronouncements of X and Y, and to reject the opinion of Z, by accepting (2). Note that the plausibility of (2) suggests the plausibility of the authorship of the manuscript by Roland of Poland, for the set (2) logically implies q by deductive implication.

Unlike deductive and inductive logic, plausibility theory tells us how we can proceed if confronted by inconsistency. In our example, we can select the most plausible subset by pruning the original inconsistent set of data. The method of plausibility screening tells us even more, however. If we look at (2) and (4), the preferred subsets in our example, we see that the two propositions, $p \supset r$ and $\lnot r$, each appear in *both* (2) and (4). In other words, no matter which of (2) or (4) we decide to accept, we are going to have $p \supset r$ and $\lnot r$ anyway. These two propositions constitute a "common denominator". And we see that the pronounce-ments of X have a certain preferred status: no matter whether we reject the sayso of Z by rejecting (2), or reject the sayso of Y by rejecting (4), we will still be accepting the sayso of X.

Plausibility screening also tells us that (2) is not consistent with (4), as a truth-table will show. Thus we must choose between (2) and (4). As we saw, it is preferable to select (2). But, in choosing (2) or (4), we would be accepting the common denominator, $\{p \supset r, \lnot r\}$.

Generally, the method of plausibility screening that we adopt consists of several steps: take the original set of pronouncements of the authorities and test them for consistency by a truth-table. If the set is inconsistent, determine all the maximal consistent subsets. Then look over the alternatives and reject those that reject any high-plausibility propositions. If there is only one left, it is the preferred set. If there remains more than one, then systematically reject the sets that tend to exclude the highest plausibility propositions until only one set is left. In the event of a tie, look to see if there is a common denominator in the tied sets. As well, you can see if there is a common denominator among any number of the maximal consistent subsets that tend to be preferable.

We could bend our example to form an illustration of a tie: suppose that Z was assigned a plausibility value of 5. Then (2) and (4) would be tied: whichever set we rejected, we would be rejecting a proposition of value 5. Either (2) or (4) is preferable to (1) or (3), but we cannot narrow the field down to one proposition. The best we can say is that you have to accept $\{p \supset r, \urcorner r\}$ because it is common to both (2) and (4).

There will be undecidable cases. If authority X says p and authority Y says $\urcorner p$, and both authorities are equally reliable, then plausibility screening will not tell us which proposition to accept. It is simply a stalemate, and we have to wait for more information.

Summary of the Method

Here is an example that illustrates, step by step, how the method works. Suppose we have the following pronouncements of three comparatively evaluated authorities, X, Y, and Z.

X (value of 9):	$p \supset q, \urcorner r$
Y (value of 7):	$q \supset r, \urcorner p$
Z (value of 2):	$p \lor q, \urcorner (p \land q)$

First, we construct a truth-table with a column representing the truth-value of each component proposition of all the propositions stated by the experts. *Second*, we scan the columns, in this case rows (1) to (8), and draw a circle around any long sequences of sets that are true. Some rows, like (7), (3), (2), and (1) below, will have true propositions in them, but these patterns of true propositions will already be included in one of the other rows. The truth-table will have four sets of circles:

	p	q	r	$p \supset q$	$\urcorner r$	$q \supset r$	$p \lor q$	$\urcorner p$	$\urcorner (p \land q)$
(1)	T	T	T	T	F	T	T	F	F
(2)	T	T	F	T	T	F	T	F	F
(3)	T	F	T	F	F	T	T	F	T
(4)	T	F	F	F	T	T	T	F	T
(5)	F	T	T	T	F	T	T	T	T
(6)	F	T	F	T	T	F	T	T	T
(7)	F	F	T	T	F	T	F	T	T
(8)	F	F	F	T	T	T	F	T	T

Third, we look at the truth-table, one row at a time, and list the propositions which we have found to be true in every circled row. We look at row (8) and see that the following columns list true propositions:

$p \supset q$, $\daleth r$, $q \supset r$, $\daleth p$, $\daleth (p \land q)$. Row (8) contains our first maximal consistent subset. Reading off the propositions for the three remaining rows with true propositions, we find four maximal consistent subsets.

(8) $\{p \supset q, \daleth r, q \supset r, \daleth p, \daleth (p \land q)\}$ Reject $p \lor q$.
(6) $\{p \supset q, \daleth r, p \lor q, \daleth p, \daleth (p \land q)\}$ Reject $q \supset r$.
(5) $\{p \supset q, q \supset r, p \lor q, \daleth p, \daleth (p \land q)\}$ Reject $\daleth r$.
(4) $\{\daleth r, q \supset r, p \lor q, \daleth (p \land q)\}$ Reject $p \supset q$, $\daleth p$.

Fourth, for each row, we list at the right the propositions of the original set that are *not* included in each maximal consistent subset.

Fifth, we scan the maximal consistent subsets as set out above, in order to evaluate their order of preference. The general rule to use when evaluating is that any set that rejects a highly plausible proposition is to be eliminated. Clearly we can eliminate (4) because it rejects $p \supset q$, which has a value of 9. Likewise (5) must be eliminated because it rejects $\daleth r$ (value 9). That leaves (6) and (8). Here the choice is also clear. Row (6) rejects $q \supset r$ (value 7), whereas (8) only rejects $p \lor q$ (value 2). On the policy that maximum plausibility is to be retained, we must eliminate (6) and therefore we accept (8) as the most suitable maximal consistent set to accept.

Sixth, we look to see if there is a "common denominator" subset that should be accepted even at the price of rejecting some sets. Looking over the maximal consistent subsets, we see that the subset $\{p \supset q, \daleth r, \daleth p, \daleth (p \land q)\}$ is common to both (6) and (8). Furthermore, we see that $\{\daleth (p \land q)\}$ is common to all four maximal consistent subsets. These "common denominators" could be tie-breakers, although in this case there is no tie – (8) stands out as the winner. Nonetheless, it is interesting to note that, despite its individual low plausibility of 2, $\daleth (p \land q)$ is acceptable because it is "carried along" in (8) and in every maximal consistent subset.

Exercise

Given the pronouncements of the authorities X, Y, and Z with a comparative plausibility rating from 1 to 10 as postulated below, conduct a plausibility screening analysis for each proposition to evaluate the preferability of the alternatives.

1.   X (value of 6): $p \supset q$, $\daleth r$
     Y (value of 5): $q \supset r$, p
2.   X (value of 8): $p \supset q$, r
     Y (value of 5): $q \supset r$, $\daleth s$
     Z (value of 2): $r \supset s$, p
3.   X (value of 9): $p \supset \daleth q$
     Y (value of 4): $r \supset \daleth p$
     Z (value of 2): r, $p \land q$

4. X (value of 8): $p \equiv q$
    Y (value of 5): $p \supset \daleth q$
    Z (value of 1): $p, q, p \wedge \daleth q$

A useful lesson from economics: Salant's Law. It is entirely appropriate to wonder why, in the expression "maximal consistent set", the word "maximal" occurs as a *noun* rather than an *adjective*. Certainly it has, in this phrase, the intended function and force of an adjective, and certainly the designation "maximally consistent set" more naturally carries the intended meaning as a consistent set of sentences, any new member of which would result in an inconsistent set. It is quite true that the use of the noun, "maximal", instead of the adjective, "maximally", is very much the received convention in logic and mathematics. This is to be regretted, since it impedes one's clear understanding with some wholly unnecessary jargon or useless linguistic eccentricity (which is what jargon is, after all). It is instructive, at this point, to turn to a bit of wisdom from Leonard Silk:

> Many economists are addicted to jawbreakers and mind-bogglers, such as *X-inefficiency, short-term profit maximization, equiproductive effort points, residual impactees, flat-top utility-effort functions, socioeconomic influence contracts, empathy utility, separability hypotheses, carte-blanche reference variables*, etc. Some economists write English as though they were translating from nineteenth-century German; they love to string nouns together. Walter Salant of the Brookings Institution has a particular antipathy towards nouns used to modify nouns to modify other nouns. When he began to run up against five-noun terms, such as *terminal traffic control program category*, Salant launched a campaign for a mandatory two-noun limit on economists.[9]

Admittedly, "maximal consistent sets" offends only the spirit and not the letter of Salant's Law; but every reader should feel completely at liberty to use the plainer (and plainly better) English phrases, "maximally consistent sets". Logicians and economists alike would do well to hearken to the good sense of Salant's Law.[10]

5. Plausible Argument

Our approach to arguments from authority has been somewhat negative and circumspect. We saw how to recognize and possibly avoid five kinds of errors in appealing to authorities. We saw that the argument from authority is not an inductive or deductive type of argument. And we saw how to rationally evaluate conflicts among authorities. But we still want to know what constitutes a correct argument from authority. What model of argument does it embody?

The key notion involved is *plausible argument*. But what is plausible argument? What are its argument links or connectives? What does it operate on, if not truth-values or probabilities?

Like inductive logic, plausibility theory is not highly developed, well-entrenched, or thoroughly investigated, as classical deductive logic has been. For all that, we present here a brief sketch of the essentials of the subject, following the pioneering work of Nicholas Rescher.

To say that a proposition is *plausible* is not to say that it is true, or even probably true, but only to say that we may presume that it is *true provisionally*. This means that it has something in its favour in the give-and-take trade-offs of the burden of proof. Nothing has come along yet to indicate that our proposition is false or unlikely, so it may be tentatively accepted, provided it does not conflict with any proposition we have already accepted in our stock of commitments. Thus the plausibility of a proposition has to do not with its *intrinsic* truth or probability, but with how *consistent* it is with other propositions that we are prepared to accept. That is why, in the previous section, we adopted the policy of retaining the highly plausible propositions in a set, even if retaining them means including some cohorts that are not very highly plausible.

If your objective was to minimize risk of error, however, it would be better to adopt another policy – you should reject all propositions of low plausibility, even if this means rejecting some highly plausible ones, too. The reason for adopting such a policy will be more clear once you understand how plausibility operates over conjunctions. The rule is:

In a conjunction of propositions, the plausibility of the conjunction is equal to the plausibility value of the *least* plausible component proposition.

If source X (value 9 on a scale of 1 to 10) asserts p, and source Y (value 2) asserts q, then the plausibility-value of $p \wedge q$ must be rated at 2. Conjunction operates something like $\wedge$ in deductive logic; in both functions, equal values tend to be preserved. If p and q are both highly plausible, then $p \wedge q$ will also be highly plausible. If p and q have a low plausibility-value, then so will $p \wedge q$ have a low plausibility-value. But "and" in probability logic is very different. Pr (p) may be high and Pr (q) may be high, but Pr $(p \wedge q)$ may be lower. For example: if p and q are independent, and Pr $(p) = {}^7/_{10}$ and Pr $(q) = {}^7/_{10}$, then Pr $(p \wedge q) = {}^{49}/_{100}$.

In plausibility logic, negation is altogether unlike negation in either deductive or inductive logic, for there is no way of determining the plausibility value of $\daleth p$ simply from the value of p. This is because, given the nature of plausibility, it is possible for both p and $\daleth p$ to be highly plausible, or for both to be of low plausibility. However, we do not want to confuse low plausibility with implausibility (negative plausibility). The difference is that implausibility is a positive reason for rejecting a proposition, whereas low plausibility is a positive – but not very strong – reason for accepting it. Basically, our analogy with inductive or deductive logic breaks down when authority X (of high value) says p whereas authority Y (likewise of high value) says $\daleth\, p$. True, they conflict; but that by itself does not determine whether we should treat p as plausible.

Another aspect where plausibility and the two logics differ is implication. For implication, the following rule obtains: *when a consistent set of premisses entails some conclusion, then the conclusion must be at least as plausible as the least plausible premiss.* In deductive logic, if a set of consistent premisses entails a conclusion that is true, that doesn't mean that at least one premiss must be true. Here are two false propositions: "If Woods wears size twelve shoes, then Woods wears glasses; Woods wears size twelve shoes." Together they entail the true proposition, "Woods wears glasses", yet both propositions are false, and they are consistent with each other. In inductive logic, as we saw, Pr $(p \wedge q)$ must be less than p and less than q, even though p and q entail $p \wedge q$. So the implication rule for plausible inference is unique.

Now that we know a little about the model of plausible argument, we are in a position to understand the core logic of the *ad verecundiam*. It is neither inductive nor deductive; it is a third type of argument, with its own distinctive characteristics.

What are some of the limitations of plausible argument? We mention three. First, as we saw, it is very often notoriously difficult to translate an expert's report into language that the layman can understand and interpret correctly. Thus (as in deductive logic) there is a translation problem. Second, appeals to expertise are *topic-sensitive*, as Condition III makes clear. But in plausibility theory, as it has been so far developed, the propositions are treated in a subject-matter-neutral way, as in deductive logic. So the fallacy of *ignoratio elenchi* is as much a threat in plausible argument as it is in deductive logic. Third, we have no way of determining whether an argument is inductive or deductive by appealing to the internal mechanisms of either deductive or inductive logic. The situation has become much more difficult now we have added a third kind of argument – plausible argument – to the list. Then, too, there were the quarrel and debate, two other types of argument. How can we keep all these diverse types of arguments from becoming hopelessly confused or misidentified? Clearly, some unifying theory must be introduced. That is the purpose of the next chapter.

Translation difficulties: an example from economics. We might, without undue offence, imagine an economist writing something along the following lines:

> Objective consideration of contemporary phenomena compels the conclusion that success or failure in competitive activities exhibits no tendency to be commensurate with innate capacity, but that a considerable element of the unpredictable must inevitably be taken into account.

In fact, however, this one-sentence passage is George Orwell's "translation" of a verse from the Bible:

> I returned, and saw under the sun, that
>   the race is not
> to the swift, nor the battle to the
>   strong, neither yet
> riches to men of understanding, nor
>   yet favour to men
> of skill; but time and chance happeneth
>   to them all.[11]

Notes

1   Augustus De Morgan, *Formal Logic* (London, Taylor and Walton, 1847), page 281.

2   Robert L. Heilbroner tells the story in his book *The Worldly Philosophers: The lives, times and ideas of the great economic thinkers*, 4th ed. (New York: Simon and Schuster, 1972), page 254. Heilbroner continues the story:

> Keynes repeated the story with relish to a friend back at Cambridge. "Why that's odd," said the friend. "Bertrand Russell was telling me just the other day that he'd also thought about going into economics. But he decided it was too easy."

3   Marlene Orton, "Science No Closer to Unravelling Mystery," *Winnipeg Free Press* (19 April 1980): 14.

4   "Sindone" is Italian for "fine cloth." The study of the Turin Shroud has produced what is now often called the "new science" of sindonology.

5   Robert Gordon Shepherd and Erich Goode, "Scientists in the Popular Press," *New Scientist* 76 (1977): 482–484.

6   C.L. Hamblin, *Fallacies* (London: Methuen, 1970), page 218.

7   Wesley Salmon, *Logic* (Englewood Cliffs: Prentice-Hall, 1963), page 64.

8   Nicholas Rescher, *Plausible Reasoning* (Assen/Amsterdam: Van Gorcum, 1976), chapter VI.

9   Silk, page 49.

10   By the way, Leonard Silk himself is guilty of a truly massive default from the spirit of Salant's Law elsewhere in the book in which he praised the law so highly. Says Silk: "The complexity of modeling a reliable economic-social-physical-biological-scientific-technological-industrial-demographic world system for forecasting purposes goes far beyond anything the MIT team [i.e. The

Club of Rome] has done". Mind you, Silk doesn't have nouns masquerading as adjectives, but he does have German syntax masquerading as English syntax, what with an eight-adjective chain!

11   George Orwell, "Politics and the English Language" in *Inside the Whale and Other Essays* (London, Penguin Books, 1955), page 149. The verse is Eccles. 9:11.

Chapter Six
Dialectic

In the preceding three chapters, we defined the *core* of argument, namely the set of propositions that makes it up (the premisses and conclusion) and the evidential relationships that obtain between these propositions. We identified three kinds of relationships – deductive, inductive, and plausible argument. We have seen that there are different kinds of argument, different kinds of premiss-conclusion relationships.

In order to evaluate any argument, it is necessary, first of all, to determine which type of argument it is. And *how* we determine the kind of argument can make a big difference to our evaluation of that argument. A correct *plausible* argument might be an incorrect *inductive* argument. And a correct *inductive* argument might be an incorrect *deductive* argument. Then, too, a correct *plausible* argument might be an incorrect *deductive* argument, and so forth. The assessment of type determines, to a great extent, the evaluation of correctness.

Confusing the three types of argument can produce many a fallacy, as we have seen. Although an argument might be a *highly plausible* appeal to authority, a deceitful detractor might effectively refute the argument – to the satisfaction of a judge or other target audience – by demonstrating, with unarguable precision, its *deductive* incorrectness. How could such a refutation be appealed or contested? Nothing in the internal mechanism of any theory – deduction, induction, or plausibility – provides a way of settling the issue or providing a defence against such a fallacy. Nor are the quarrel or the debate any help, for they are only too open to toleration of this sort of refutation.

So we need another approach to argument, one that is not open to the excesses and fallacies that can occur in the debate or quarrel, but one that is more attuned to argumentative interchange than is the theory of deduction, induction or plausibility. Deduction, induction, and plausibility concentrate exclusively on the core of the propositions and three

110

evidentiary relationships among them. We need to take a more panoramic view of argument, in order to co-ordinate these three elements and fit them into the cut and thrust of objection and rebuttal. We turn now to *dialectic*.[1]

1. Basic Rules of Dialectic

The basic idea behind the theory of rational dialectic can be stated simply.

Dialectic is the art or practise of examining statements logically, as by question and answer, to establish validity.

Unlike the debate and the trial, but like a quarrel, dialectical argument is one that the participants themselves preside over and judge. Unlike the quarrel, dialectic is carefully governed by rules and is deliberately designed to discover the truth of the matter at hand. Like the criminal trial, there is an emphasis on evidence and what it allows a reasonable person to infer or conclude.

Centuries ago, Aristotle proposed the following rules or procedures for dialectic arguments:

1. There is one questioner and one answerer, although substitution is permitted.
2. The answerer is obliged to defend some thesis on a certain topic, and his defence takes the form of responding to probing, critical questions by the questioner. The questions are asked and answered one by one.
3. For his part, the questioner must ask clear and straightforward questions, including what in law are called leading questions. Otherwise, the answerer may refuse to respond, and demand that the question be clarified or re-expressed.
4. The answerer must give only his own honest conviction in reply, not what he thinks will merely improve his chances of defence. If, however, his answer contradicts his previous answers, he must amend one or the other until the contradiction is removed.
5. If the answerer cannot, through ignorance, give a reply, it is the questioner's job to formulate new questions that will overcome that ignorance.
6. Answers may not be postponed.
7. The exchange concludes if the answerer is refuted or if it is clear to both participants that the questioner's refutation will not succeed.[2]

Essential to the effective functioning of dialectic is a willingness of the participants to display objectivity, fairness, and goodwill. The unwary or naïve dialectician risks being cut to ribbons by an unscrupulous opponent, or by an opponent who is prepared to substitute the style and the objectives of mere debate for those of dialectical inquiry.[3] In any real-life dialectical situation, a dialectician must always be prepared to detect such abuses and to withdraw from the fray. In a dialectical discussion, it is always better to keep one's powder dry and one's procedures intact, and to be ready to fight again another day.

Aristotle's model of dialectic is not graven in stone. It is not the only possible or reasonable conception of dialectic. In fact, it would seem that Aristotle's rules are slightly too restrictive to serve as an appropriately general conception of rational inquiry by co-operative question and answer. Accordingly, we will suggest an amended and somewhat relaxed version of Aristotle's model, as follows.

(1) How Many Questioners and Answerers? Aristotle determined that there should be just one of each. It would seem that the principles that motivated Aristotle's rule were:

    a) that the number of participants involved should not *overload* the discussion;
    b) that the ratio of questioners to answerers should be sufficiently in balance as to allow for a *fair* discussion.

These are admirable principles, which may be accepted without hesitation. However, when you think about it, Aristotle's "one-on-one" application of the principles will not always be the best application. For example, if Bill is highly trained in bio-medical ethics, and Sue is very well versed in bio-medicine but has no training in ethics, a one-on-one dialectical exchange between Bill and Sue would hardly be fair. Suppose Sue were joined by Hank, who knows no biology or medicine but whose training in ethics is quite impressive. Would this not be a *fairer* contest with a greater chance of reaching a competent conclusion? Thus, the first requirement of dialectic is this:

    (i) The total number of participants, and the relative numbers on each side of the question, should be small enough and sufficiently balanced that the exchange is manageable and fair.

(2) What Are the Main Tasks of the Participants? For Aristotle, there is a two-fold dynamic in a dialectical inquiry. A thesis is:

    a) *tested* by questions designed to refute it, if possible; and
    b) *defended* by the answering of those questions as truthfully as the answerer can.

It is not clear whether Aristotle also requires that questioners test only those theses that are already believed to be false, and answerers defend only those theses that they already believe to be true. On reflection, it would seem that dialectic need not be constrained in this fashion. If the two-fold dynamic of dialectic is effective, then the refutation of the thesis, or the failure to refute it, might be expected to occur irrespective of the initial convictions of the participants. There is no particular reason that, in the dialectical give-and-take, a participant might not surprise himself with an outcome in which a disbeliever is made into a believer, or a believer into a disbeliever. Thus the second rule of dialectic is:

> (ii) Answerers must defend a thesis by responding to critical queries of questioners whose aim it is to refute the thesis. Questioners need not initially disbelieve, nor answerers believe, the thesis at hand.

(3) What Kind of Questions Can Be Asked, and How Are They to Be Asked? Aristotle is quite right to require that questions be intelligible, and quite right to allow the respondent to reject them if they are not. On the other hand, it is not necessary (or desirable) that the dialectician's questions be *transparent*; they need not clearly indicate the questioner's entire strategy of refutation. Clarity is required; naïveté and predictability are not. The third rule, then, is:

> (iii) The questioner must ask clear and straightforward questions; otherwise the answerer may refuse them or demand that they be clarified or re-expressed.

(4) In What Manner Should the Answerer Make His Responses? He must answer *honestly* and *consistently*. That is, he must tell the truth as best he sees it. In answering a question, he must not contradict any previous answer. If he does contradict an earlier response, Aristotle requires him to remove the inconsistency before the discussion can continue. He must either change the answer to the current question or he must change the answer previously given.

Here, too, Aristotle is a bit severe. We can think of two quite different ways of handling contradictory answers, either of which may be appropriate in certain circumstances. First: perhaps the contradiction concerns an *inessential detail*; perhaps, when confronted with his two conflicting answers, the questioner will not know which reply to stick with. At this point, Aristotle recommends breaking off the dialectical exchange. But it would seem more reasonable that both participants agree to *ignore the contradiction*. Essentially, the questioner would withdraw both questions. Second: perhaps the contradiction concerns a point of importance, and hence cannot be ignored. But, again, perhaps the answerer honestly does not know which of the two conflicting answers to give up. Rather than abandon the discussion, it might be more reasonable to "bracket" the conflicting answers, to put them on hold, as it were. The participants can return to the issue once the discussion has taken on a more substantial shape, at a time when the participants might feel more confident about dealing with the contradiction. When quarantining a contradiction, remember that the dialectic can continue to unfold only if its development does not depend upon a resolution of that contradiction. Thus the fourth rule of the dialectic is:

> (iv) The answerer must reply truthfully and consistently. However, if he does contradict an earlier answer, then (I) he must withdraw one or

other answer; or (II) with the consent of the questioner, the contradiction can be ignored as inessential; or (III) again with the consent of the questioner, the contradiction can be quarantined to be returned to later, and the discussion can continue if its development does not depend on a prior resolution of the contradiction.

Thus dialectic can deal with the circumstantial *ad hominem*, a fallacy that could not be resolved by the quarrel. Given that a participant has been shown, by his opponent's questions, to be caught in an inconsistency between his present recommendation and some circumstance of his own practice, he may resolve the contradiction by following rule (I), (II), or (III). Thus the participant's recommendation is not necessarily destroyed by his contradiction, for the questioner must give him the option of resolving it, according to the fourth rule of dialectic.

Exercise

Go back to the examples of the circumstantial *ad hominem* studied in Chapter One (page 11). Show how they could be resolved, in a dialectical dialogue, through applications of the fourth rule of dialectic.

(5) How to Handle an Answerer's Ignorance[1] Sometimes a respondent cannot reply because he simply does not know what answer to give. Aristotle requires the questioner to take the leading rôle in overcoming the ignorance that blocks the reply. Aristotle further stipulates that the removal of the respondent's ignorance be restricted to a series of *supplementary leading questions*, and that these questions be organized into a kind of *Socratic form of teaching*.

Evidently, Aristotle is trying to avoid or minimize the possibility that an answerer will merely accept a questioner's confident assertion concerning the matter about which the answerer is ignorant. Such a confident assertion may not be true. And, true or not, the answerer's acceptance of it may not be supported by an *adequate understanding* of what he accepts. His acceptance may therefore run afoul of the dialectical rule that all answers be honestly given.

But although the Socratic method is one good instrument for replacing ignorance with genuine understanding, it is not the *only* way of dealing with ignorance. Aristotle is right to imply that an answerer should not merely take the word of the other participant concerning matters about which he is himself ignorant. But it does not follow that an answerer might not *sometimes* take a questioner's word. Remember, the whole structure of dialectic depends on the integrity and goodwill of the participants.

Aristotle is too skeptical when he assumes that the assertions of a questioner are always to be distrusted – especially when those assertions are designed to help an answerer overcome his ignorance. It is also over-skeptical to say that, whenever an answerer accepts the assertion of a questioner in such circumstances, he is always accepting something on blind faith, something he does not understand. We need only consult our own every-day dialectical experiences to recognize that this extreme skepticism is out of place in many situations. Accordingly, the fifth rule of dialectic is:

(v)   When his ignorance makes it impossible for an answerer to reply, the questioner must take a leading rôle in removing that ignorance. He may avail himself of the Socratic technique of question and answer to try to elicit from the answerer a new understanding, which ultimately will defeat the answerer's ignorance. If there is not reason to think that the questioner would be untruthful or would not know the answer to his own question, then it is permissible for the questioner to volunteer to the answerer what he takes to be the correct answer. Then, if the answerer understands what he has been told and genuinely believes that it is a correct answer, he is at liberty to adopt it as his own answer.

(6)   May Answers be Postponed? Aristotle flatly says no, and we can certainly see why. If an answer can always be postponed or ducked, then there is no guarantee that the dialectical exchange will ever get anywhere. It becomes possible for the answerer to resort to the simple and dishonest expedient of postponing answers to difficult and probing questions; the onus is on the questioner to defeat *any such attempt at refutation*. But as our discussion of Rule (iv) suggests, it does seem that in certain circumstances answers may, by mutual consent, be postponed to be dealt with at a later time. Accordingly, the sixth rule of dialectic is:

(vi)   Answers may not be postponed, except for reasonable cause and by mutual consent.

(7)   What Brings the Debate to a Close? Aristotle would seem to have had the right view of things. Accordingly, the seventh rule of dialectic is:

(vii)   The dialectic exchange terminates when either the answerer is refuted or when it has become clear to both participants that the questioner will not succeed in his refutation.

Even these amended rules of dialectic only partially and somewhat abstractly fit the rich and often complex texture of actual dialectical inquiry. In most dialectical exchanges, each participant will play the rôle of *both* questioner and answerer at various stages of the discussion, and with respect to various points at issue. It is not by any means unusual for Sue to begin as questioner and continue in that capacity for a while, yet for Bill to sooner or later say something like this: "Now, Sue, let *me* ask *you* something. From the tone of your questions up until now, you would seem to believe that so and so, but is it not the case that such and such?" At this juncture, there has been an *exchange of dialectical rôles*. In actual dialectical situations, a number of difficult questions arise; with respect to some of those questions, one participant might feel more confident than another. On the face of it, then, there is no good reason that this shifting network of confidence should not be reflected in shifts of dialectical initiative and dialectical function, to a certain extent. However, if such shifts are to be allowed to occur in our model, they can be allowed only by mutual consent. Accordingly, we have an eighth rule of dialectic:

> (viii) By mutual consent, dialectical shifts can be allowed.

We could no doubt offer further rules of dialectical procedure. But for the time being, the eight rules we have already presented convey a sufficiently developed notion of rational inquiry by co-operative question and answer to serve our present purposes. Subsequently we shall expand upon the uses of dialectic and raise still more questions about its logical structure. For the present, we can say that the logical description of dialectic is to be found in our eight rules, which we here repeat for the convenience of the reader.

THE EIGHT BASIC RULES OF DIALECTIC

1. The total number of participants, and the relative numbers on each side of the question, should be small enough and sufficiently balanced that the exchange is manageable and fair.
2. Answerers must defend a thesis by responding to critical queries of questioners whose aim it is to refute the thesis. Questioners need not initially disbelieve, nor answerers believe, the thesis at hand.
3. The questioner must ask clear and straightforward questions; otherwise the answerer may refuse them or demand that they be clarified or re-expressed.
4. The answerer must reply truthfully and consistently. However, if he does contradict an earlier answer, then (I) he must withdraw one or other answer; or (II) with the consent of the questioner, the contradiction can be ignored as inessential; or, (III) again with the consent of the questioner, the contradiction can be quarantined to be returned to later, and the discussion can continue if its development does not depend on a prior resolution of the contradiction.
5. When his ignorance makes it impossible for an answerer to reply, the questioner must take a leading rôle in overcoming that ignorance. He may avail himself of the Socratic teaching technique of question and answer to try to elicit from the answerer a new understanding, which ultimately will defeat the answerer's ignorance. If there is not reason to think that the questioner would be untruthful or would not know the answer to his own question, it is permissible for the questioner to

volunteer to the answerer what he takes to be the correct answer. Then, if the answerer understands what he has been told and genuinely believes that it is a correct answer, he is at liberty to adopt it as his own answer.

6. Answers may not be postponed, except for reasonable cause and by mutual consent.
7. The dialectic exchange terminates when either the answerer is refuted or when it has become clear to both participants that the questioner will not succeed in his refutation.
8. By mutual consent, dialectical shifts can be allowed.

2. A Practical Exercise in Dialectic: the Discussion Group

The best way to gain real familiarity with the give-and-take of dialectical argument is to practise it yourself, with a partner or in a group interaction. Accordingly we propose, as a major project for this chapter, that the student arrange, with one or more class-members, a small group discussion; you should plan on one hour each week for four sessions, or more if needed. Pick a controversial ethical issue as a topic. Then, by a combination of research and group discussion, identify and formulate plausible arguments for *both* sides of the question. The group must write up a dialectical account of the arguments, giving several arguments for each side and also outlining objections to each argument. Finally, a reply to each objection must be offered.[4]

Procedure

1) Form a group of three or four members. If you have difficulty finding partners, contact the instructor.
2) At the first meeting, canvass possible topic areas and choose the one you propose to investigate. At the second meeting settle on some common readings as a basis for discussion. Allow for a fairly broad-based discussion until the issues in the selected area become familiar.

Topics

Choose a topic that can be characterized as a "contemporary moral issue." Suggestions are:

(1) Abortion;
(2) The legal enforcement of morality;
(3) Egalitarianism and meritocracy;
(4) Censorship;
(5) Euthanasia;
(6) Capital punishment.

Assume a controversy in each area: some will hold that the practice or action in question is morally right and others that it is morally wrong. Your task is to make explicit the arguments that may be offered in defending or refuting these conflicting positions. As you examine the issue, you may find it essential to make distinctions and to narrow the scope of the question, to give it a manageable form. For instance, in discussing euthanasia, you might want to focus only on cases of mercy-killing of terminally ill patients who request it.

Format

The first problem is formulating the issue so that the question can be framed in a clear and specific way. The following series of progressively more specific statements, offered by Maner, give an idea of what might need to be done to avoid side issues.[5] The phrase in italics sharpens the focus of each successive formulation. In our example, we are discussing the acceptability of cohabitation.

    a)   Is it true that cohabitation should be accepted in our society?

    b)   Is it true that cohabitation should be accepted in our society *as a preliminary to marriage?*

    c)   Is it true that cohabitation should be accepted in our society as a preliminary to marriage *for adults?*

    d)   Is it true that cohabitation should be accepted in our society as a preliminary to marriage for adults *who are healthy, self-supporting, and knowledgeable about contraception?*

    e)   Is it true that cohabitation should be accepted in our society as a preliminary to marriage for *mature persons of legal age* who are healthy, self-supporting, and knowledgeable about contraception?

The clarification of crucial terms deserves careful attention. Try to avoid the possibility that ambiguity might send you on a chase after secondary or irrelevant considerations. Ideally, formulation or statement of the issue might include this clarification; for example, (e) above might be viewed as making more precise what is meant by "adults" in (c). Another way to clarify is to append an explanatory paragraph, in which any possibly confusing terms used in the formulation of the issue are defined. All of this must be achieved without editing out the interesting issues!

Two components should be added to your report:

1)   Background information
    a)   List names of participants.
    b)   Indicate frequency of meetings, what was accomplished at each session, and the total time spent on the project.
2)   Recording and reporting
    a)   One person should be designated "secretary" to record the arguments. Or, if possible the sessions may be recorded on tapes, for use by the participants in writing up their individual reports.
    b)   Each individual report will consist of a "fictionalized" dialetic modelling the main arguments, objections and replies, after the pattern of the report in Appendix I.

Suggestions and Reminders

1. Co-operative endeavour doesn't always work as smoothly as a given individual may want it to. First, there is the problem of dealing with conflicting opinions. Fortunately, this assignment is designed to enable you to take advantage of the divergency of opinions so far as the dialectic is concerned. Your individual report gives you, in the end, the opportunity to disavow any part of the dialetical interchange that you think detracts from its strengths.

2. Prepare for the meetings by doing the required reading and thinking about the issues involved. The responsibility falls to each of you to formulate the questions you think most need investigating. Bring notes with comments and questions, to the group meeting.

3. Arrange for everyone to sit in a circle so that each person can see everyone else.

4. The degree of verbal participation will undoubtedly vary among members of the group. Those who find it easy to talk should take care not to dominate the discussion unduly. If you find it difficult to speak up, consider this your best opportunity to transcend that difficulty.

5. Each person who comes prepared has as much right as the next to determine the course of the discussion. You have a right to have your contribution acknowledged, though it might not be accepted.

6. You must try to discipline yourself to stick as much as possible to the topic and not indulge yourself by mentioning whatever happens to come to mind. Listen carefully. Ask yourself whether your contribution continues or adds to the question of the moment, or whether it is likely to be a distraction. But don't be too reticent: one must always consider the possibility that an initially unlikely suggestion might turn out to be relevant and significant when pursued a bit. Also, there comes a point in a discussion when a change of direction is necessary.

7. Since your topic is likely to be based on information that takes different positions, don't be afraid to engage in controversy. Your first reaction might be to avoid the expression of differences because such a discussion seems unpleasant or threatening. But consider the real possibility that you can come to understand your own position more fully when it is explored against other alternatives.

8. You can approach the give-and-take of these sessions with a view to self-knowledge rather than trying to subjugate others to accept your views. You don't need to express your opinions in a hostile way and thereby discourage others from taking part. Try not to make the discussion group a win-or-lose debating forum.

9. While disagreement should not be suppressed, it is important to try to separate the rejection of reasons for a view, or even of the view expressed by another, from attack *against the person*. (It is, admittedly, a nice question how far this is possible when we are dealing with very fundamental beliefs.)

10. A good way to start each session is for several people to attempt to characterize how the main issue or problem is formulated in the

selection(s) being examined. Then ask questions about the question or problem. Is it clearly presented? Is it open to more than one interpretation? Are there presuppositions or assumptions that need, themselves, to be questioned? Is there a frame of reference you need to adopt in order to appreciate the force of the question?

11. Now you might proceed to examine the different positions taken. Positions not yet mentioned might occur to you; if so, introduce them. Suggestions should be offered for clarifying and evaluating the reasons offered for and against the different positions. Invite the opinions of as many group members as possible over the course of the whole session.

12. You will probably not have time to cover all facets of each issue that is raised. It is not necessary – indeed, it will rarely be possible – to close the consideration of any item of discussion with uniform agreement. But ideally, the value of shared investigation of difficult issues will be realized as each participant receives feedback for his or her expressed thoughts and finds fresh suggestions for developing his or her own thinking still further, through the challenges presented by the opinions of others.

13. To collect information on your topic, go to the library and find a collection of essays on the issue. Circulate these readings amongst the participants so that everyone has a chance to gain familiarity with some of the basic arguments employed by writers on your topic.

14. Consult the sample Discussion Report on Cannibalism in Appendix I (page 251) as a model.

3. Dialectic: Its Evaluation

Dialectic is certainly a benign form of argument, and a true friend of reason. Its procedures are designed to promote objectivity and fairness, and its structure allows for a range of outcomes or results that enhances our capacity, however modestly, to get at the truth of things. Here are some examples.

(1.) Refutation in the Strong Sense

One outcome of the dialectical exchange is that a given thesis is refuted in what we call *the strong sense*. To say that a thesis or proposition is refuted in the strong sense is to say that it is shown to be *false*. Now, in general, if one knows that a given proposition, p, is false, one also knows that its opposite, not-p, is true. So refutation in the strong sense is one means of discovering the truth.

(2.) Refutation in the Weak Sense

A thesis may be refuted in what we call *the weak sense* when the discussion shows that the answerer has clearly insufficient grounds for holding the thesis in question. However, it does not follow, from the fact that I do not have enough evidence to show my thesis to be true, that my thesis is false. For all I know, it may be *true*, yet I do not have sufficient evidence to be able to *show* that it is true. In this kind of situation, we can say that the thesis is refuted in the sense that its defence is demonstrably inadequate. It can be seen, therefore, that the refutation in the weak sense of a given proposition does not entitle one to claim that the proposition is false and that its opposite is true. However, knowing that one's thesis has been refuted in the weak sense is something very much worth knowing. One comes to understand that, irrespective of whether that thesis is true or false, one's *grounds for holding* the thesis are inadequate; therefore, one's claim to *know* the thesis is deficient.

By the way, we now briefly introduce another of our fallacies, the *argumentum ad ignorantiam*; that is, the sort of argument directed "to (a person's) ignorance." In one rather broad example, the fallacy goes something like this:

> A cure for cancer hasn't been found.
> Therefore, cancer knows no cure.

The upshot of this bad argument is that, since a cancer cure is not yet known, it doesn't exist. In fact, we can now see that the fallacy is quite nicely described this way:

The ad ignorantiam is the fallacy of thinking that refutations in the weak sense are (also) refutations in the strong sense.

Exercise

Decide whether an *ad ignorantiam* fallacy occurs.

1. Nobody has ever disproved the existence of ghosts. Therefore ghosts exist.
2. Nobody has ever proved that we have free will. Therefore, every action is determined by heredity or environment.
3. The conjecture that low-level radiation is linked with cancer and other diseases has never been demonstrated by clear scientific evidence. So low-level radiation is safe.

(3.) Stalemate

A third outcome of a dialectical argument is that the participants will concede, sooner or later, that the questioner's critical attack is not going to succeed in refuting the answerer's thesis in either the strong or weak sense. In his failure to achieve a refutation, the questioner cannot say that he knows the thesis to be false; nor can he say that he knows the answerer's defence of it to be inadequate. On the other hand, even though the refutation of this thesis has not succeeded, the answerer is not automatically entitled to assume that his thesis is therefore true, or that his grounds for holding that thesis are entirely adequate. *It may be that the thesis is false and that the questioner has not been astute enough to penetrate the answerer's defence.* Thus it is *possible* that a false proposition will not be exposed under dialectical examination. But the very fact that a proposition holds up to such examination constitutes *some degree of evidence* that the proposition has been capably and rationally defended, and also that it might be true. Certainly if a thesis holds up under *repeated* dialectical examination *from all quarters*, its truth will appear more and more likely.

A Point to Remember: mind you, dialectic is not the perfect form of argument, and on the whole it is no more successful than its participants are informed, astute, reasonable, and fair. It is, then, something of a deficiency of dialectic that it admits of a certain weakness in execution.

For example, dialectic is not well-served if it matches a "heavy-weight" against a "light-weight". If the answerer is not competent to offer a good defence of his thesis in the face of sharp and probing critical examination, then it is fairly certain that, *irrespective of the merits of the thesis*, the dialectical exchange will result in a refutation. There is one exception to note:

Sometimes dialectic is used not as a means of testing a thesis for its truthfulness or lack of it; instead, it is used as a *teaching device*. In such *didactic* uses, dialectic very deliberately assigns the rôle of questioner to a well-experienced teacher and the rôle of answerer to a relatively inexperienced pupil. Then, by skillfully questioning, the questioner brings his pupil, step by step, into an enlarged understanding or state of knowledge about the proposition at hand. From the point of view of a contest of intellects, there is not much to be said about this use of dialectic. But from the point of view of effective teaching of what is already known, by the teacher at least, it can be a very effective application of dialectic.

Similarly, dialectic will not serve its fundamental purpose at all well if it matches two light-weights against one another. Exchanges between the merely ignorant or the generally incompetent are not likely to throw much light on the thesis in question; they are certainly not likely to constitute an adequate test of any thesis. At the other extreme, it is possible to reserve dialectical functions exclusively for the well-matched *experts*. But, as we have already seen, there are problems with such exclusiveness. For one thing, the notions *of expertise* and *the expert's authority* are not always as well behaved as we might wish. Some appeals to the authority of expertise are fallacious; that is, they constitute the traditional fallacy of the *ad verecundiam*. On the other hand, it is also clear that sometimes judgments and decisions cannot properly be made except with the assistance of expert opinion. Nevertheless, in restricting dialectic to exchanges between experts there is certainly the risk of committing the fallacy of *ad verecundiam*.

Secondly, whether we like it or not, the business of life, which involves a multitude of judgments to be made and decisions to be weighed, is a business for us all – for experts and plain folk alike. No human being can escape the urgent importance of making up his own mind and making his own way in a complicated and often dangerous world. Like it or not, therefore, it falls to us all to do the best we can to make our decisions and to form our judgements and adjust our strategies on the basis of what has the best chance of being true and the best chance of winning our confidence rationally. Unless we are to surrender our liberty and our capacity for autonomous action, it is up to us all to make our best estimate of what the truth is, and to determine our best courses of action. It would be a pity, then, to deny to the ordinary man and woman or to the not especially gifted person the benefits of participation in the clarifications of dialectic. This is not to say that each of us is capable of a *good* dialectic examination of *every* question in the world. But there are questions that lie at the very heart of human activity, questions that we must all address and make up our minds about. We would stand a better chance of dealing with these questions effectively were we to employ the mechanisms of dialectic.

There is a third problem. We have already mentioned the susceptibility of certain kinds of dialectical arguments to the fallacy of the illegitimate appeal to authority or expertise. In this connection, we should also mention *two other fallacies*. Bearing in mind that dialectic is an inquiry generated by a process of question and answer, it should come as no surprise that some of the fallacies to which dialectic is prey are *fallacies dealing with questions*. Here are two examples.

The Fallacy of the Complex Question

Not every question is a proper question to ask. For example, to take a very obvious and well-known illustration, it is perfectly unacceptable to ask the ordinary husband, "Have you stopped beating your wife?" What is wrong? Somehow the question contains an *illicit* and *unsupported assumption*, namely that the person to whom the question is directed *has been* a wife-beater. The effect of this would seem to be that, *whichever way* the answerer responds, whether he says yes or no, he is undone and humiliated by the question.

Begging the Question

If some questions are not fit questions to ask in a dialectical exchange, some answers are not good answers to offer there, either. Here is another obvious example; so obvious, in fact, that it scarcely fools anyone (but for now we ignore the subtleties of such fallacies). The answerer might begin a dialectical argument by presenting the thesis that he wishes to defend, the thesis, let us say, that God exists. The questioner then asks whether the answerer has any good grounds for holding to this thesis. The answerer replies that he has, namely, that he has found the appropriate authoritative assurances of God's existence in the Bible. The questioner then asks why it is that the answerer takes the authority of the Bible to be decisive; the answerer replies that it is because the Bible tells the authoritatively infallible truth on the matter. The questioner now asks why the answerer takes the Bible to tell the infallible truth; the answerer replies that it is because the Bible is the word of God and that God, by nature, is truthful. But in this last remark, it is straightforwardly *assumed* that God exists; it would seem that the whole force of the answerer's replies concerning the existence of God depend upon the *prior assumption* that God exists. Logicians call this blemish of reasoning *begging the question*. This fallacy will be dealt with in detail in Chapter Eight.

We now turn to an examination of the fallacy of complex question and some other associated question-and-answer fallacies of dialectic.

4. The Fallacy of Complex Question

A *whether-question* poses a number of alternatives, of which the answerer is supposed to select one. For example: "Is she wearing the red dress or the green dress?" Each of the alternatives is called a *direct answer*. Any statement implied by a direct answer is called a *presupposition* of the whether-question. Take, for example, the whether-question, "Is she wearing the red dress and the purple hat, or is she wearing the green dress and the puce shoes?" In effect, the question poses an alternation of two conjunctions. That is, it says: $(R \wedge H) \vee (G \wedge S)$? Thus the direct answers are $R \wedge H$ and $G \wedge S$. An example of a presupposition would be R, because R is implied by the direct answer $R \wedge H$. So now we see that questions can have presuppositions. Thus a question may not always be as innocent as questions often seem. Merely *asking* a question may not seem to be in any way harmful, risky, or fallacious. After all, to just ask, we may think, is not to assert anything positive, or to make or require commitments. But what if the question has a presupposition? In a dialectical interchange, presuppositions can be dangerous. Just asking may not be so innocent; it might be a subtle way of disguising an aggressive shift of the burden of proof.

To sort out the relative riskiness of presuppositions, let us divide propositions into three classes: the logically necessary ones, the logically inconsistent ones, and the logically contingent ones. We have already met up with inconsistent propositions. They are the ones, like $p \wedge \lceil p$, that *cannot be true*, for every row of the final column of their truth-table is false. The logically necessary ones are those like $p \vee \lceil p$, where every row of the truth-table is true. The contingent ones are those that are neither necessary nor inconsistent. In the truth-table of a contingent proposition, there is a mixture of truth and not-truth in the final column.

Exercise

Determine which of the following propositions are necessary, which contingent, and which inconsistent:
(a) $(p \vee q) \wedge (p \vee \lceil q)$
(b) $(p \wedge q) \vee \lceil (p \wedge q)$
(c) $(p \wedge q) \wedge (p \wedge \lceil q)$
(d) $(p \supset q) \supset (q \supset p)$
(e) $(\lceil p \vee q) \supset (p \wedge \lceil q)$

Every whether-question presupposes that at least one of its direct answers is true. This idea, that at least one of its direct answers is true, is called *the* presupposition of the question. A whether-question is called *safe* if its presupposition is logically necessary, risky if it is not safe. For example, "Is she wearing the red dress or not?" is a safe question because its presupposition, $R \vee \lceil R$, is logically necessary. *Yes-no questions* are always safe because their presupposition consists of a pair of contradictory alternatives. For example: the presupposition of the question, "Is snow white?" is, "Snow is white or snow is not white." The presupposition is logically necessary.

Exercise

For each of the following questions, give an example of a direct answer and a presupposition. Then determine *the* presupposition of the question, and show whether the question is safe or risky.

1. Are you a pacifist?
2. Are you a pacifist or not?
3. Are you a pacifist or a loyalist?
4. Are you a pacifist or a non-pacifist warmonger?
5. Are you a pacifist or if not then a warmonger?

The exercise shows that, even if a question is safe, it may not fairly represent a range of reasonable alternatives. Suppose I ask you whether a zebra is black or white. This question is not safe; even worse, it is not a fair question, because it does not offer enough alternatives. The alternatives should be these: a zebra is black but not white; white but not black; both black and white; neither black nor white. That is, *the* presupposition of the question should not be B ∨ W; instead, if it is to fairly represent all the relevant possibilities, it should be (B ∧ ⅂ W) ∨ (W ∧ ⅂ B) ∨ (B ∧ W) ∨ (⅂ B ∧ ⅂ W). So reformulated, the question is not only safe; it represents a fair range of alternative direct answers.

The following are actual titles of works published by reputable historians:

> Napoleon III: Enlightened Statesman or Proto-Fascist?
> The Abolitionists: Reformers or Fanatics?
> Plato: Totalitarian or Democrat?
> The Dred Scott Decision: Law or Politics?
> Ancient Science: Metaphysical or Observational?
> Feudalism: Cause or Cure of Anarchy?

These questions suffer from a wide variety of faults of shallowness and simple-mindedness. But whatever else their faults, they are splendid examples of the Black-and-White Fallacy.[6]

Now we are in a position to appreciate, a little more clearly, how the fallacy of many questions arises. Let us look again at our example: "Have you stopped beating your wife?" This question is stated in the yes-no form and therefore fools us into thinking that it should be safe. But, in fact, it has a substantive presupposition. Let W = "You have a wife and have beaten her"; let S = "You have stopped beating her". The question says: (W ∧ S) ∨ (W ∧ ⅂ S)? But this is truth-functionally equivalent to the ordinary statement, W. So the question is risky. If the presupposition W is actually false, it is impossible to give a true direct answer to the question, because W appears on both sides of the disjunction, (W ∧ S) ∨ (W ∧ ⅂ S). Thus the only sensible answer is to "correct" the question by pointing out the falsity of W. The fallacy occurs when a question that is actually risky, and moreover has a false presupposition, is put in the guise of a safe "yes-no" question.

But not all questions that have substantive presuppositions are fallacious. "Is she wearing the red dress or the purple hat?" is not fallacious, because its presupposition is not concealed. Not all risky questions that have a *false* presupposition and are in the form of a "yes-no" question (or another form of question that appears safe) are fallacious. Even if it might be true that you have a wife whom you have beaten, putting the question in such an aggressive way leaves you no opportunity to attempt to deny it. Dialectically, such foreclosure is not to be tolerated, for it does not enable the answerer to respond with denial or defence.

The question, "Have you stopped beating your wife?" in effect permits only two possible (simple and direct) answers, *yes* or *no*. But either answer implies that the answerer *has* beaten his wife. A simple flow chart demonstrates this point.

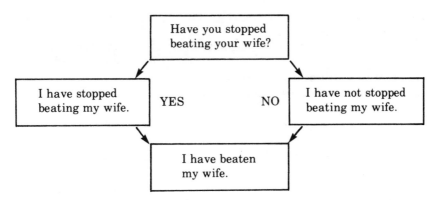

The answerer is trapped. No matter which alternative he chooses, *yes* or *no*, he is committed to the same unwelcome implication. The *fallacy* is: what appears to be a genuine choice of alternatives is, in reality, a trap.

If a question is safe, it is "logical" to expect that it has no unwelcome presupposition. And, given its form, we expect the wife-beating question to be safe. That is our initial, reasonable perception of it, because of its "yes-no" form. But this expectation clashes with the question's underlying analysis. Not only is it risky, contrary to expectation – it even has an unwelcome presupposition – one that the answerer is forced into, no matter which way he answers.

Dialectic provides a solution to these fallacies. A participant who is confronted with a question from his opponent is allowed to ask that the question be reformulated if it contains a presupposition that the answerer does not accept or that is unwelcome to him. The answerer may ask that the question be *divided* into smaller units, so that he can respond separately to each unit.

The proper response to the wife-beating question, dialectically, is for the answerer to request that it be broken down into two questions: (1) Do you have a wife you have beaten? (2) If you have a wife you have beaten, have you stopped beating her? To say no to the first question is to excuse the answerer from responding to the second, and that puts an end to the inquiry. The proper response to the zebra question is for the answerer to request that the question be reformulated so that it is separated out, thus representing a greater range of possible alternatives.

There is nothing fallacious about asking a question with an unwelcome presupposition, one that the answerer does not accept or will not like. Nor is there anything wrong with asking a question that has a complex presupposition, say of the form, "If p then q", or "Both p and q." But if a question combines several presuppositions, then the answerer must have the option of responding to each presupposition separately. To attempt to force him to answer in such a way that this option is removed is to commit the fallacy of complex question.

Exercise

What, if anything, is fallacious about the following questions?
1. What did you use to wipe your fingerprints from the gun?
2. Are you still a heavy drinker?
3. Do you feel guilty when you are involved in your criminal activities?
4. Don't you feel we should either get out of this foreign entanglement or show some military muscle and smash the guerilla hideouts with tactical nuclear weapons?
5. Do you jog, or are you sedentary?
6. If you're such an anglophile, why don't you move to England? [Hint: could *ad hominem* be involved?]
7. Have you exhibited questionable judgment and unstable behaviour in the past, or are these cognitive disabilities part of a recent pattern of deterioration?
8. If ethics is not a science and is therefore purely subjective, why do you try to defend your guilty behaviour when you know yourself that you are being completely illogical?
9. How can you explain the amazing success of astrological predictions in the past?

5. Summary

Rarely will a real-life argument, one in which there is a serious exchange of contending views, conform to just one of the models of argument that we have been examining. If you pay careful attention to the logical structure of actual argument, you will see that, at various stages, various different models tend to predominate. An argument might *begin* as a case of dialectic at its best. Yet, after an hour, it might have taken on rather more of the flavour of the skilled and icily polite debate; after another hour, it could very well have disintegrated into a screaming match. When you yourself are engaged in an argument, you should always be alert to the possibility of an unannounced shift from an argument of a given model to an argument of another model. Plainly, if I do not detect that you have abandoned the dialectical model for the debating model, I could very well cease to be an effective opponent for you, and I might end up making concessions that I need not make.

We now see that there exists a spectrum of various kinds of arguments; the spectrum is defined by two extremes: the quarrel and the deduction.

In general, the nearer an argument is in this spectrum to the quarrel the more it will be a personal argument; and the nearer it is to the deduction, the more impersonal it will tend to be.

This is not to say that people *cannot* engage in deductive or inductive exchanges with one another, nor is it to say that these deductive or inductive exchanges cannot touch on very personal aspects of their lives. Rather, the point is this:

A quarrel *cannot help* but be a personal argument, whereas a pure deduction or an induction *need not* involve any personal elements whatever.

Dialectic is the most general approach to argument, for it incorporates deductive, inductive, and plausible arguments; at the same time, it provides a framework for questions and answers, for the interpersonal give-and-take of argument and refutation. So dialectic is partly impersonal, but also partly personal.

Because dialectic can at times include the personal, some logicians are inclined to think of *deduction* as the paradigm and acme of objective reasoning. We do not entirely share this point of view. Instead, we assume that *dialectic* provides more general and at the same time more practical standards of what constitutes good argument. After all, dialectic aims at the improvement of the understanding, the elimination of error, and the pursuit of truth. Its mechanisms and procedures are designed to be fair and objective. Besides, dialectic can incorporate into its own development the rigour and power of deduction, where this is applicable, and also good strategy. It can also incorporate induction and plausibility. Then, too, dialectic transcends some of the limits of deduction, induction, and plausibility. If a certain truth is inaccessible to proof or disproof because of the limitations of deduction, induction, or plausibility, there is no reason to think that an effective dialectical exchange might not reach the desired conclusion.

And, finally, dialectic is admirable because it is so practical a mode of reasoning. Many real problems do not admit of exclusively deductive, inductive, or plausibilistic settlement, and rarely is an effective and reasonable solution ever found by quarreling. We have also seen that the debate and trial by jury are, in actual, every-day practice, only tangentially the friend of reason and truth. Dialectic empowers you to press an opponent in ways that require him to defend his thesis and to find for it an adequate justification; you, in turn, are at liberty to probe and scrutinize the thesis for its soundness. The dialectical method, therefore, has a flexibility that enables participants to stick to what is relevant and to pursue what is likely to be helpful. The dialectical argument has all the capacity that it needs to grow and to shape itself in ways that skillfully put to the test the object of its investigation.

Because it incorporates deductive and inductive logic, dialectic can deal with deductive and inductive fallacies. And, by incorporating plausibility theory as one of its types of argument, dialectic copes with the *ad verecundiam*. Because it allows us to resolve or handle contradictions, dialectic provides the best framework for dealing with the *ad hominem* argument. We have also seen how dialectic enables us to recognize and analyse the *ad ignorantiam*.

The models of argument we studied previous to dialectic, collectively had several problems. They did not enable us to tell whether what we were confronted with was an argument or not. That is, they did not enable us to locate or identify arguments. For example, the *ad baculum* and *ad populum* were often difficult to handle because we could not determine whether we were really faced with an argument. And, if we did have an argument, the other models did not enable us to tell what *type* of argument we faced.

These difficulties are straightforwardly resolved in dialectic. In a dialectical interchange, all participants are required to identify the type of argument that they put forward if any participant expresses doubt on this score. Dialectic demands that the moves in argument must follow convention: questions must be clearly distinguished from assertions; both must be clearly distinguished from arguments. For example, these are four characteristic moves in a dialectical sequence.

(1) Do you accept *p*?
(2) Why do you accept *p*? [Why do you not accept *p*?]
(3) I accept *p*. [I do not accept *p*].
(4) I accept *p* on the (deductive, inductive, plausible) argument that *q* and if *q* then *p*.

The pattern of interchange allowed in the usual dialectical sequence is basically question-response. Thus, to initiate a sequence, the first participant may choose a move in the form of (1) or (2). If he or she inappropriately begins with a move of the form (3) or (4), immediately the outcome is decided in favour of the opponent. Similarly, if the question is of form (1), response of form (4) is inappropriate. Rather, the response must be of form (3). If the question is of form (2), then either a response of form (4) is required, or a request to reformulate the question is in order [fallacy of complex question].

Thus dialectic always enables us to distinguish between arguments and non-arguments, and to clearly mark the type of argument that is involved. In numerous significant ways, then, dialectic is the most general model of argument, and incorporates or supersedes the previous five models of argument.

Notes

1   The word "dialectic" derives from the old French *dialectique*, which, in turn, can be traced to the Greek *dialektike*, meaning "discussion by question and answer". Dialectic should not be confused with the philosophical doctrine of *dialectical materialism* of Marx and Engels, which is a theory of historical development.

2   Aristotle, *Topics*, Book VIII; see also C.L. Hamblin, *Fallacies* (London: Methuen Co. Ltd., 1970), pp. 61-62.

3   Remember poor Cleinias, in the passage quoted from Plato's *Euthyphro*.

4   The statement of procedures and format of group discussions is based on materials for classroom use designed by Professor V.Y. Shimizu. We are grateful to Professor Shimizu for offering us the benefits of his experiences with these techniques.

5   Walter Maner, *Practicum Manual*, unpublished manuscript.

6   For an interesting discussion of the widespread susceptibility of historical writings to this fallacy, see: David Hackett Fischer, *Historians' Fallacies* (New York and Evanston: Harper & Row, 1970).

Chapter Seven
Arguing in a Circle

We have already noted that an argument may, on the one hand, fail its intended objectives and yet, on the other, succeed with respect to unintended objectives. Similarly, one and the same argument may be a very good argument, judged by certain criteria, and a perfectly dreadful argument, a real catastrophe, judged by other criteria.

It is of fundamental importance to appreciate that rarely is an argument either just good or just bad. Frequently, one and the same argument can be a good and successful argument of one sort, and yet not a good or successful argument of another sort. A clear example of what we mean is demonstrated by this simple argument:

(1) All the change in my pocket is made up of nickels.
(2) Therefore, all the change in my pocket is made up of nickels.

Now it is perfectly clear – to take an obvious example – that this argument will not do

i) as a refutation of its premiss.

How could it succeed as a refutation of its premiss? The conclusion is not manifestly false; as well, it merely reaffirms what the premiss already says.

Nor will this argument do

ii) as an induction of its conclusion.

We have spoken before of arguments in which the premisses increase the *likelihood* of the conclusion without strictly proving it. Such arguments may be said to be more or less successful *inductions* of their conclusions (as contrasted with deductions of those conclusions). In our example, the argument cannot be construed as even a marginally successful induction of its conclusion, since the conclusion and its premiss are the very same proposition. Consequently, the likelihood of the conclusion will be *exactly the same* as the likelihood of the premiss; it will not to any degree be *increased* by that premiss.

There is no difficulty in seeing our sample argument as a perfectly good *pure deduction* of its conclusion. As we learned, an argument is pure deduction if its premisses entails its conclusion. That is, it is pure deduction if its premisses are such that, should they be true, the conclusion is guaranteed also to be true. If you look at our sample argument, you can see that premiss and conclusion are one and the same proposition. Therefore it could not happen that the premiss is true, yet the conclusion is false. *Every proposition entails itself.*

Nevertheless, nearly everyone would agree that this good deduction is a bad argument. One does not have to be talented in logic to see at once that this argument commits the fallacy of *arguing in a circle*. This fallacy is sometimes called the *fallacy of petitio principii*, and we shall devote some considerable care, in this chapter, to the examination and evaluation of its structure. For now, we meet it in a form so obvious that it would not fool anyone in an actual, every-day situation. In this perfectly obvious form, it seems clear that arguing in a circle essentially involves using the conclusion as one of its own premisses. So arguing in circles would seem to be a kind of *argumentative pulling yourself up by your own bootstraps*. From one point of view, an argument can be a perfectly good deduction of its conclusion; from another point of view, that same argument can be a dreadful example of *petitio principii*.

We have mentioned John Stuart Mill in connection with our evaluation of the debate and the trial by jury. Mill also had some interesting things to say about deduction. Perhaps the most interesting claim that Mill ever made about deduction – and certainly the most controversial – was this:

All deductions commit the fallacy of *petitio principii*.

Mill's point can be grasped by considering another example of a deductive argument:

> All men are mortal.
> Socrates is a man.
> Therefore, Socrates is mortal.

Here, we can all agree, is an unimpeachable deduction. Yet Mill complains that the conclusion is already contained in, or implicit in, the premisses; in asserting the premiss, you are already, implicitly at least, asserting the conclusion. Since the premisses already say what the conclusion says, the argument is circular.

And, since this second example is a perfect paradigm of an impeccable deduction, it follows that all deductive arguments (all correct ones, that is) are circular and commit the fallacy of *petitio principii*.

This is a startling claim to make. If Mill is correct, we would have to concede that every *correct mathematical proof*, even though a perfectly competent deduction, is *fallacious*! Before we evaluate Mill's claim, let us try to get a better idea of the fallacy of *petitio principii*.

1. Two Models of *Petitio*

The fallacy of *petitio principii* is thought to occur when an arguer somehow illicitly assumes the very proposition he is supposed to be proving.[1] Suppose Bob, a skeptical agnostic, and Lester, an evangelical fundamentalist, are conducting a theological disputation of a familiar sort.

Bob:	What reasonable grounds could we have for believing that God exists?
Lester:	It says so, quite clearly, in the Bible.
Bob:	Admittedly. But how can we be assured that what the Bible asserts is, in this respect, accurate?
Lester:	The Bible is the word of God. We can trust it completely, especially in a case like this, where its message is obvious.
Bob:	Hold on a second!

Bob may suspect that Lester is indeed assuming the very proposition he has set out to prove. We can see this if we break Lester's argument down into its steps or stages.

> p: Conclusion: God exists.
> q: Premiss for p: "God exists" is written in the Bible.
> r: Premiss for q: The Bible is the word of God.

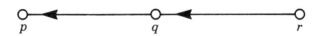

$$p \qquad\qquad q \qquad\qquad r$$

But the problem for Bob is that r would seem to assume p. That is, the evidence to support r could hardly exclude p. Taking the Bible as a seriously authoritative source of the "word of God" presupposes the existence of God. Thus it would seem that, whatever evidence Lester might give to support r, that evidence can hardly escape being based, at least in part, on p. But how can r be based on p? As our diagram shows, p is based on q, and q is based on r, and therefore p is indirectly based on r.

Thus r is supposed to be based on p, *and* p is apparently supposed to be based on r. The argument seems to be pulling itself up by its own bootstraps! Or, to put it another way, the argumentative links are forced into an interlocking closed pattern, in this case a triangle:

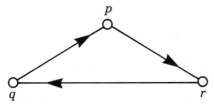

Small wonder that Bob hesitates, fearing logical entrapment. That closed-in feeling he no doubt has may be connected with his suspicion that somehow the argument has gone in a circle. And so it has.

Of course, Bob might have other reservations about Lester's argument. It seems, at least in part, to be an argument from authority, the authority of scripture, and therefore it might raise suspicions of the *ad verecundiam* fallacy. But at least arguments from authority can sometimes be good arguments. Bob is likely most concerned with the structure of the argument as a whole. That structure is a closed chain, even if each of its separate links might, by itself, be a legitimate inferential move. Each link might indeed be questioned. But the argument *as a whole* may be questioned. Bob should concern himself with the soundness of the argument's structure.

Both historical and contemporary textbooks comment on two aspects of *petitio*.

---

1. According to the *dependency conception*, an argument is thought to be circular where the conclusion is presupposed by some premiss, in the sense that the premiss actually depends on the conclusion for part of the evidence that backs it.

Exponents of the dependency conception often express their judgment of what is *fallacious* about the *petitio* in terms of what can or can't be inferred from a conclusion. By these lights, an argument is *non*-circular only if one may know that each premiss is true without having to infer that truth from the conclusion or from some statement that can only be known by inference from the conclusion. It is easy to see how the dependency conception is exemplified in our example of the theological disputation: there, it seemed that the premiss, r, could only be known to be true by an inference that would depend on p, the alleged conclusion of the argument.

---

2. According to the *equivalence conception*, an argument is circular where the conclusion is tacitly or explicitly assumed as one of the premisses; that is, where the conclusion is equivalent or even identical to a premiss.

Here is Irving Copi's discussion of the equivalence conception:

> If one assumes as a premiss for his argument the very conclusion he intends to prove, the fallacy committed is that of *petitio principii*, or begging the question. If the proposition to be established is formulated in exactly the same words both as premiss and as conclusion, the mistake would be so glaring as to deceive no one. Often, however, two formulations can be sufficiently different to obscure the fact that one and the same proposition occurs both as premiss and conclusion.[2]

An example will help: "Unlimited freedom of speech must usually be advantageous to the state, because it is in the best interest of the community that each individual should enjoy unlimited liberty of expression." If you think about this argument a bit, you will see that the premiss (the part after "because") is nothing but the conclusion, repeated perhaps in somewhat different terms, but not different enough to add any new or significant evidential backing to the conclusion. After all, the phrase "unlimited freedom of speech" amounts to the same thing, it would seem, as "unlimited liberty of expression". And "in the best interest of the community" is a phrase that amounts to pretty much the same thing as the expression "advantageous to the state". There is hardly a significant enough difference to indicate that the argument amounts to anything much more than p, $\therefore p$.

Perhaps, in the end, the dependency and equivalence conceptions are really the same fallacy. (Only a further explanation of the key notions of "equivalence" and "dependency" would reveal what real practical or theoretical difference there is between the conceptions.) Geometrically, the dependency conception expresses the idea that the premisses, p, and the conclusion, c, of a chain of arguments are points on the circumference of a closed figure. That is, we can follow a line of inference from p to c:

If the argument is circular, we can also follow a line of inference from c to p, because p must somehow depend on c in a truly circular chain of inferences.

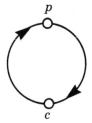

Thus, in the dependency conception, *p* and *c* are like points on the circumference of a closed space (like a circle). But if you pick a point on a closed space, you will necessarily eventually come back to that very point. The equivalence conception expresses just this idea, that an argument is a *petitio* if a premiss takes us back to that very same proposition that is the conclusion.

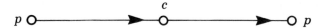

In other words, the argument is circular if we start at some point in it and must arrive back at that very same point.

Problem for Analysis

Lester and Bob are again locked in a theological dispute. Bob has expressed his deep doubts that the traditional Christian concept of God can escape the charge of anthropomorphism.[3] He asks: "Why should our concept of God be based on a model evidently utilizing essentially human attributes as a basis? Why, for example, should God be thought to be benevolent?" Lester finds it hard to select a good argument that would be appropriate to the situation because he does not share Bob's doubts. But he knows that Bob is a science major and a keen student of mathematical logic, so he decides to propose an argument that is incontestably valid in formal logic: "God has *all* the virtues, therefore God is benevolent." Bob starts to look worried but dubious. Is Lester's argument circular? If so, which conception of circularity does it represent?

Exercise

Indicate whether a fallacy is committed by the following responses; if there is a fallacy, what is its nature?

1. Response of one politician to the question of why another lost his seat: "He didn't get enough votes."
2. Bob committed suicide because he had a death wish.
3. Margaret Laurence is a better writer than Ross MacDonald. Why? Because people of literary taste and judgment prefer

Laurence. But how does one identify the class of person who has literary taste and judgment of the required sort? Why, they are the very people who can be identified by showing their preference for Laurence over MacDonald.

2. Locating the Fallacy of *Petitio*

Augustus De Morgan was a staunch defender of the uses of formal logic; he is known for his contributions to the subject in the nineteenth century. De Morgan also had some original and very useful thoughts on the subject of informal fallacies. In keeping with his championship of deductive formal logic, De Morgan preferred a classically deductive version of the equivalence conception as an account of the *petitio:*

> ... strictly speaking, there is no formal *petitio* except when the very proposition to be proved, and not a mere synonyme of it, is assumed.[4]

But his contemporary, Alfred Sidgwick, was quite justifiably skeptical about how adequate an account of *petitio* could be given that confined itself to purely formal cases such as $p, \therefore p$ and $p \wedge q, \therefore p$. These cases are indeed plausibly circular, but tend to be trivial. Can more complex cases somehow be reduced to these formulae? Sidgwick was doubtful:

> ... nothing appears to be really gained by restricting the name to so small a compass as this and there is no doubt that such a restriction would be very much at variance with the popular acceptation of the term.[5]

Thus a general question is raised about the exact limits of classical logic in the study of the *informal* fallacy of *petitio*.

In evaluating these limits, we might first observe a point of considerable importance:

A given deductive argument form can have an instance that is non-circular – and yet the very same instance can also be circular in different dialectical circumstances.

Let us consider an example in the form of the *disjunctive syllogism*. In our example, Alvin and Calvin, two alpinists, have departed together on an expedition to Mt. Baldy. We now proceed to consider two hypothetical sets of evidentiary circumstances for the same argument.

The Argument

> Either Alvin or Calvin is missing.
> Alvin is not missing.
> Therefore, Calvin is missing.

First Dialogue (no circularity here)

Bob: I just heard a report on the radio that one of the two alpinists has apparently fallen into a crevice and has not been found.

Sue: Did they determine which one it was?

Bob: No. But wait – I can see somebody coming down the trail. It's definitely Alvin. Good grief, that means Calvin is missing!

Sue: (very reluctant to concede that it could be Calvin who fell) But can we really be sure that Calvin is missing?

Bob: Well (states disjunctive syllogism).

Second Dialogue (circularity here)

Bob: I have been learning in logic class today about something called the disjunctive syllogism.

Sue: I took that last year. What else is new?

Bob: Well, this form of inference is interesting stuff. I can use it to prove that Calvin is missing, provided only we assume that Alvin over there is not missing.

Sue: It sounds to me like you're missing something yourself.

Bob: No, wait – you remember the Rule of Addition, that, given p, you can always derive $p \lor q$ for any q you like?

Sue: Yes, that always struck me as a trifle fishy. But I learned to live with it when the instructor told me that in deductive formal logic, a disjunction "says less" than either of its disjuncts, in the sense of being a weaker statement. So in deductive logic you can derive $p \lor q$ from p, even if p and q seem unrelated.

Bob: Very philosophical. Anyway, using that rule of addition, I can prove that Calvin is missing by disjunctive syllogism (he utters the disjunctive syllogism). Now you're probably wondering how I can prove the first premiss. Easy – using the rule of addition, "Either Alvin or Calvin is missing" is a logical consequence of "Calvin is missing".

Sue: You're not playing fair. "Calvin is missing" is what you're supposed to be proving. That's a *petitio*.

Bob: A what?

> *Alvin:* I think Bob is a little confused. That *petitio* is
> so gross it wouldn't fool anyone.
> *Sue:* He is going to have trouble with logic.

What our two dialogues indicate is that *whether or not an argument is circular is not determined by its deductive logical form*. In both of our dialogues, that argument is in the form of the disjunctive syllogism. In both dialogues, therefore, the argument is deductively correct. However, in the first dialogue the argument is, to all appearances, non-circular, and indeed a perfectly good, reasonable and unfallacious argument. The argument in the second dialogue is a howler – such a bad *petitio* would hardly be expected to fool anyone who is moderately attentive, or at least anyone who is not philosophically confused about formal logic. Therefore:

An instance of the disjunctive syllogism can be circular or not depending on the dialectical circumstances in which it is advanced.

Our discoveries so far suggest that the logic of the *petitio* is not to be sought exclusively in the deductive logical form of the argument.

3. Mill's Puzzler

The nineteenth century philosopher John Stuart Mill argued for the shocking thesis that *all* deductive reasoning is circular. The thesis is shocking because it implies that all deductive logic is infected with the fallacy of *petitio*.

Mill was a champion of inductive logic. One can therefore appreciate why he might have been inclined to belittle deductive logic. But his reasoning went deeper than mere personal preference or whim. This can be seen if we reflect on the example he offers for analysis. His argument is not only deductively correct; historically, it has usually been taken as the very essence of deductive correctness (if such correctness could be conveyed in a single argument).

> All men are mortal.
> Plato is a man.
> Therefore, Plato is mortal.

In his *System of Logic*, Mill argued that this argument is circular because the first, or "major", premiss depends on the conclusion. That is, we cannot be assured that all men are mortal "unless we are already certain of the mortality of every individual man." If it is doubtful that Plato (or anyone, for that matter) is mortal, then it is at least as doubtful that all men are mortal. Mill seems to be thinking of the dependency *petitio*. He argues that we cannot accept that the major premiss is true unless we already accept that the conclusion is true. By similar reasoning, Mill argued, we can take *any* deductively correct argument and show that some premiss cannot be known to be true unless the conclusion is already known to be true. Therefore, all deductively correct arguments beg the question. So much for deductive logic!

What have logicians had to say about this shocking accusation? Morris Cohen and Ernest Nagel argue that Mill has set his standards too high.[6] In order to know that a universal statement is true, they say, it is not necessary to know that *each* and *every* instance of it is true. Such a requirement would seem to demand absolute certainty in matters of fact. But in experimental science, it must be accepted that no universal hypothesis can be known with absolute certainty. The number of possible confirming or disproving instances is open-ended and cannot be closed if the hypothesis is to be applicable to future (and consequently possibly undetermined) instances.

Of course, we could think of a universal statement like "All men are mortal" as a simple enumeration of a finite number of instances. It would then become a finite conjunction: "Man_1 is mortal $\land$ man_2 is mortal $\land$ . . . $\land$ man $_i$ is mortal (for finite i)". In effect, this conjunction would be a list of all past men and all presently living men: John Woods is mortal *and* Douglas Walton is mortal *and* Alfred Sidgwick is mortal *and* Gene Autrey is mortal *and* et cetera, et cetera. And somewhere in the list, presuming it is complete, we will find the statement "Plato is mortal". If we *did* think of the statement "All men are mortal" in this finite way, then the argument Mill cites would indeed be circular. In fact, it might be an *equivalence petitio*, because the conclusion, "Plato is mortal", would be one of the sub-premisses of which the major premiss is conjunctively composed. And certainly it would be a *dependency petitio*, because the long conjunction could not be true if one of its conjuncts, "Plato is mortal", were not true (as we will remember from the truth-functional definition of conjunction).

Granted, then, that the argument would be circular *if* the major premiss could be treated as a finite list of instances. But Cohen and Nagel object; they say that it would not be treated as finite in the sort of setting in which such an argument might occur in experimental science. They conclude that Mill has adopted an over-simplified and naïve view of the logic of experimental science, which cannot be restricted to finite generalizations.

But let us give Mill a chance to defend himself. He might grant that we do not need to know that each and every possible instance of a universal hypothesis is true in order to accept the hypothesis as a scientifically useful generalization. But surely even the most enlightened exponent of scientific method would agree that, if a counter-instance of a generalization is found, that generalization must be given up (or at least modified, so that in effect a more limited generalization is proposed). That is, Mill might still argue that if I know the statement "Plato is mortal" is false, then I know the statement "All men are mortal" is false. So, according to Mill, the premiss still depends on the conclusion. The premiss cannot be true unless the conclusion is true.

But a critic might question whether Mill has the right interpretation of the dependency relation. Surely, as long as we have no serious doubts about the statement "Plato is mortal", the body of historical and biological evidence independent of this one particular instance is quite strong enough to lend considerable support to the major premiss. Therefore, in the absence of some evidence of Plato's immortality, the argument carries considerable weight as a non-circular marshalling of evidence for its conclusion. In short, Mill's accusation cannot be resolved conclusively until we can achieve a clearer grasp of the dependency relation on which the dependency conception of *petitio* is built.

We seem to be caught in a stalemate. Mill's argument is no superficial puzzler, but a deep difficulty that forces us to reflect more carefully on the structure of the *petitio* itself as a form of argument. It is hard to see how an account of the *petitio* could be forthcoming in classical deductive logic if all correct deductive arguments are circular, or if it is variable whether correctness correlates with circularity. Following Mill's lead, we might be ready now to turn to a dialectical approach.

Something like Mill's argument had been known even in ancient times.[7] And De Morgan, no doubt aware of its deeper implications, was ready to concede that it was "ingenious". Nevertheless, De Morgan offered a way of bypassing the difficulties the argument implies while at the same time refuting the argument itself. He replied with what we call *De Morgan's Deadly Retort:*

> The whole objection tacitly assumes the superfluity of the minor [i.e. second premiss]; that is, tacitly assumes we know Plato to be a man, as soon as we know him to be Plato.[8]

We are assuming that Plato is not a dog, for example, or a computer program. De Morgan grants that if the conclusion is false then the major (first) premiss must also be false – but only on the assumption that the minor (second) premiss is true. The one premiss seems to beg the question if taken by itself. But if we consider the argument as a unified whole, it is not circular.

De Morgan then generalizes that the most favourable account of *petitio* refers to what is assumed in *one* premiss:

> The most fallacious *pair* of premisses, though expressly constructed to form a certain conclusion, without the least reference to their truth, would not be assuming the question or *an* equivalent.[9]

De Morgan illuminatingly connects the phenomenon of *petitio* to the question of the *number* of premisses of an argument. Like many significant developments of logic, this line of thought was known to the ancient Greeks.

In ancient times, the Stoics disputed whether one-premissed *argument* can exist. Antipater of Tarsus (150 BC *fl.*) led the group that stood for the existence of one-premissed arguments. An example: "It is day, therefore it is night."[10] The opposition maintained that this argument should be amplified to read "If it is day, it is night. It is day. Therefore it is night." Evidently, the ancients thought that in order to have a legitimate, full-fledged *argument*, two independent premisses must be involved.

Aristotle also requires that an argument have at least two premisses.[11] The traditional logic of *all* and *some* was constructed by Aristotle, and was the dominant theory of logic until the twentieth century. According to Aristotelian Logic, an argument is required to have *exactly* two (independent) premisses, no more and no less. Nowadays we consider this limitation arbitrary, but in the last century it was natural to think of formal logic as dealing exclusively with two-premissed arguments of a certain pattern, which were called syllogisms.

A *classical syllogism* is a sequence of two premisses and one conclusion. The syllogism contains three *terms*; each term must occur in exactly two of the three propositions. A term refers to a class of things. Here is a correct syllogism:

> All flies are insects.
> All insects are cold-blooded.
> Therefore all flies are cold-blooded.

The three terms are "flies", "insects" and "cold-blooded [things]." Notice that the syllogism is made up from these terms in a particular way; the pattern is characteristic of the traditional logic of *all* and *some*. Exactly two terms appear in the conclusion; each occurs once. The second term of the conclusion ("cold-blooded") is called the *major term*. The first term of the conclusion ("flies") is the *minor term*. The two premisses are constructed in such a way that the major term occurs once in one of them and the minor term occurs once in the other. A third term, the *middle term* ("insects"), occurs once in each premiss but does not occur at all in the conclusion. Each of the three terms occurs exactly twice in the argument, and each proposition of the argument contains exactly two distinct terms. To conform to the pattern, each syllogism must contain exactly two premisses and each premiss must be logically independent of the other.

Since syllogisms are always multi-premissed, we can state De Morgan's thesis on petitio succinctly in the traditional terminology of syllogistic as follows: *No syllogism is circular.* But syllogistic logic is archaic; it is much more interesting to attempt to state the thesis in terms adaptable to modern logic:

De Morgan's Thesis

No argument with more than one premiss, and no superfluous premisses, is circular.

We may now define the notion of a *superfluous premiss* in a correct argument. A *premiss* in a correct argument is *superfluous* if, and only if, the argument is still correct if that premiss is deleted. Take, for example, an argument of the form "$p \supset q, p$, therefore q." Is the second premiss, p, superfluous? Let us try deleting it. The result is $p \supset q$, therefore p, which is incorrect. In the same manner, we learn that the first premiss is non-superfluous. Hence each premiss is logically *independent of* the other, since neither is a deductive consequence of its partner. This form is called *modus ponens*, and all arguments of this form are many-premissed arguments without superfluous premisses. Now let's look at the argument form "p, q, therefore p." We can see that the first premiss is not superfluous. But the second one is, for if we delete q we are left with "p, therefore p" which, we remember, is a deductively correct argument form.

Now we can get down to the heart of the matter. We remember that the disjunctive syllogism is a form of argument that has substitution instances that are sometimes circular and sometimes not. But the disjunctive syllogism is many-premissed, and none of the premisses is superfluous. Take the disjunctive syllogism $p \lor q, \neg p$, therefore q. If we delete $p \lor q$, we are left with $\neg p$, therefore q, which is invalid. If we delete $\neg p$ we are left with $p \lor q$, therefore q, which is also invalid. Neither premiss is superfluous. Here, then, is an apparent counter-example to De Morgan's Thesis. We know that instances of the disjunctive syllogism *can*, under the right evidential conditions, commit a *petitio*. But contrary to De Morgan's Thesis, the disjunctive syllogism is a many-premissed argument and neither premiss is superfluous. Consequently De Morgan's Thesis might not seem to be applicable to modern logic.

4. De Morgan's Defence

But it is not likely that a defender of De Morgan's classical point of view would give up easily, even when confronted with arguments like "the two alpinists". Presumably, he would try to show that two different arguments are involved in the circular and non-circular cases, and that the difference between them is somehow reflected in their respective deductively logical forms. De Morgan, or his defender, might argue as follows. Yes, both instances of the two-alpinist arguments share the form of argument characteristically known as the disjunctive syllogism.

$$p \lor q$$
$$\neg p$$
$$\overline{}$$
$$q$$

This description of form contains all the parts of the non-circular argument. But it does not completely describe the circular argument. Remember that Bob argued for the disjunctive premiss on the basis of the conclusion. Thus there is actually another part to the circular argument, namely

$$q$$
$$\overline{}$$
$$p \lor q$$

If we look at the argument as a whole, we can see that it really consists of *two stages*. We need a new form of argument to reflect this piggy-back feature:

$$q$$
$$\overline{}$$
$$p \lor q$$
$$\neg p$$
$$\overline{}$$
$$q$$

First we have the premiss q, then the intermediate conclusion, $p \lor q$. Then the intermediate conclusion is used as premiss for the next stage of the argument, which is a disjunctive syllogism. But this representation of the form of the argument reveals its circularity *in a deductively formal manner*. The statement q appears as both a *premiss in* and a *conclusion of* the larger argument, the argument that contains both stages. In short, it seems that De Morgan is vindicated. If the argument really is a *petitio*, then we can show that the conclusion is exactly identical to a premiss and not "a mere synonyme", just as De Morgan said. A circular argument might appear to have the *form* of the disjunctive syllogism. But deeper analysis of its logical form will reveal that the conclusion is being smuggled in as a premiss; the argument is revealed as a classical equivalence *petitio* of a purely deductive sort. Deeper analysis of the alpinists example reveals two distinct arguments, each of which has a different logical form, an apparent look-alike.

De Morgan's Defence can be stated in a *general* way, and we can combine it interlockingly with De Morgan's preference for the equivalence conception of *petitio*. That is, we can sometimes explore the logical form of an argument by discovering that one premiss of it is really the conclusion of another argument:

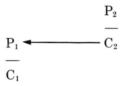

A *second stage* of the argument may emerge. And this process of regression could be followed along through numerous steps to any stage i, where i is finite.

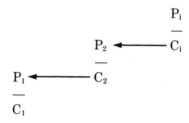

In arguments, it happens often that a critic or skeptic may not be satisfied with your premiss, and asks for some additional premiss that could be used as a base for reasonably accepting your original premiss. This process is called *dialectical regression*, and it follows the argument back through each successive stage. In effect, the larger, major argument is made up, in chain-like fashion, from a number of smaller arguments or links. Presumably, the De Morgan defender would suggest that arguments (and their logical forms) must be viewed from a "macroscopic" perspective as well as from the usual, or "microscopic", perspective we are accustomed to. In a macroscopic representation of the form of the argument, we will find a clear line of demarcation between the circular and non-circular arguments if we push the analytical regression back far enough to encompass the circle. So we now see that the disjunctive syllogism, as in the alpinists example, is not an exception to De Morgan's Thesis; the Thesis can be applied to modern logic. And it would seem that we can deal with *petitio* exclusively in terms of classical logical forms. So concludes the defender of De Morgan's Defence.

De Morgan's Defence shows a number of interesting new aspects of the concept of an argument. Notably, it shows how arguments can be linked together to form *chains of argument*. P.T. Geach calls the logical form of an argument its *schema*; when schemata are linked together in this fashion, he calls the resulting chains *themata*.[12] For Geach, themata are not arguments as such; instead, they are distinct logical units in their own right, just as an argument is made up of propositions but is not itself a proposition. It would seem likely that for Geach, arguments themselves are not circular, but the themata that join them together can form a circle. These observations give us some insight into the fact that many intervening argument steps are sometimes required to close the circle.

By introducing the notion of step-like sequences of arguments, does De Morgan's Defence successfully obviate the need to appeal to epistemic concepts in the analysis of circularity? The answer here depends on how one interprets the arrows we used in representing the generalized Defence. Suppose we have a really circular argument. If we follow the step-by-step process of isolating, each time, the proposition a certain premiss is *based on*, we will eventually come back to the conclusion of the original argument we started with. So the key notion is that a premiss in a given argument is *based on* a premiss in some other argument that "backs it up" evidentially.

We think the notion of "based on" (represented by the arrow in the diagrams on page 146, which the De Morgan procedure uses, must be essentially dialectical in nature; further, we believe the notion can be seen properly only from a dialectical perspective. In the regressive De Morgan procedure, at each step we are driven behind a premiss to extract its *evidentiary basis* in a repeated sequence of what might be called *retroductive analysis*. Usually an argument is thought of as a forward evidential step from premisses to conclusion. But we can also look at any argument in the reverse way; we can move backwards from some proposition, or from the conclusion, to its evidentiary basis. If we push backwards far enough, we will find the conclusion of the argument again at some eventual stage in the progressive retroductive analysis *if the argument is circular*, that is, if there is *petitio*. At least, so it appears from the point of view of the equivalence conception of the *petitio*.

But the very same phenomenon can also be interpreted as an instance of the dependency conception. Let us look again at the original one-step argument. We naturally think of the conclusion as evidentially dependent on the premisses. But by a series of retroductive steps, we might find that a premiss is dependent evidentially on some *previous* evidential base, which depends on yet *another* proposition. And so on and so on until, by a series of steps, we arrive back at the original conclusion. We have now shown, step-by-step, that the original premiss ultimately depends on the conclusion that we ultimately arrived at. *Hence the dependency circle*. The conclusion depends on a premiss but, if the argument is a *petitio*, the premiss also turns out to depend on the conclusion. In short, once we understand the arrow (→) that represents the notion of a step or stage in a sequence of determinations of the relation of evidentiary basis in a retroductive analysis, we will have grasped the fundamental building-block of the *petitio*.

But let us look back at the classical syllogistic example cited by John Stuart Mill.

(A) All men are mortal.
(B) Plato is a man.
(C) Therefore, Plato is mortal.

We can now see that more than a purely classical account of this characteristic relation must be involved. The upshot of our discussion above was that the syllogism might or might not be circular, depending on how the first premiss is to be verified by the arguer.

If the arguer has some way of verifying the first premiss by an appeal to general biological laws that involve no specific reference to the particular individual "Plato", it seems hard to see how the argument really is circular.

On the other hand, if the arguer were thinking of that first premiss as a finite enumeration of *all* men, then statement (C) ("Plato is mortal") would need to be a part of his evidentiary basis. Or he might otherwise indicate or admit that statement (C) is indeed part of his evidentiary basis. Then the charge of *petitio* is very much to the point. Let's look at the argument again, and examine it deductively. We see that (A) taken with (B) entails (C). But to say that (A) taken with (B) entails (C) *is not by itself sufficient* to sanction a well-founded allegation of *petitio*. It is the *dialectical* factor of what the arguer chooses as his basis of argument that is crucial. Or, to put it another way: it is not the purely deductive logical form itself that is circular or not-circular: the *petitio* is relative to the *dialectical circumstances* of that form of argument. The critical question resides in the evidentiary basis of the first premiss question. The presence or absence of *petitio* can be determined only by examining *how* the arguer infers the first premiss or *why* he accepts this premiss as true. Consequently, this process of retroductive analysis so characteristic of *petitio* which was ingeniously uncovered by De Morgan only serves to highlight the need to go beyond the classical deductive logic and to think of argument in the dialectical setting.

Exercise

1. Find an example of a deductively valid argument; then find a dialectical setting that makes the argument appear circular. How could Mill have used your argument to provide evidence for his thesis that all deductive logic is circular? How could De Morgan attempt to refute the argument you have found?

2. Find any superfluous premisses in the following arguments:
   (a) $p \supset q, \, \neg q \supset \neg p, \, p$, therefore q
   (b) $p \vee q, \, \neg p, \, \neg q$, therefore q
   (c) $p \supset q, \, \neg p, \, p \vee q$, therefore $\neg q$.
3. Construct an example of a dialogue to illustrate dialectical regression.

5. Circles in Dialectical Games

Aristotle used the phrase "begging the question" to describe the fallacy of arguing in a circle. The phrase refers to a dialectical game: one party is required to defend a certain proposition, and that party asks his opponent to grant, as a premiss, the very proposition he is supposed to defend. It seems safe to assume that the opponent (the second party) does not accept the proposition that the first party is required to prove to him – and herein lies the fallacy. Given that the second party's original dialectical position is *not to accept* the proposition – in other words, given that the stance he has taken towards it is that of non-acceptance – it seems pointless to ask the second party to accept that very proposition as a premiss. So it seems reasonable that dialectic would have some apparatus for dealing with the problem.

Still, as we all know, sometimes there is nothing wrong with asking an opponent to accept a premiss that he may not presently accept. So what is really wrong with begging the question? The answer is *that nothing need be wrong with it* – except that it involves a certain redundancy or repetition. Suppose Woods asks "Why p?" and Walton replies "Because p." If he wishes to query the premiss, Woods must ask "Why p?" again. The process could go on and on – and on! By forcing Woods to keep asking the same question, Walton could create a kind of filibuster. He couldn't achieve a successful resolution of the dialectical interchange, but he could prevent Woods from doing so; thus he might achieve a kind of stalemate.

There are ways Walton could achieve the same effect; we call them *circle games*. We'll make each game as simple as possible dialectically. There will be two participants, Black and White, and White always moves first. White is the questioner, and Black tries to give premisses that will justify the proposition queried, at each move, by White. That is Black's objective.

I.	White	Black
	1. Why p?	Because p.
	2. Why p?	Because p.
	et cetera	

II.		White	Black
	1.	Why p?	Because q.
	2.	Why q?	Because p.
		et cetera	

III.		White	Black
	1.	Why p?	Because q.
	2.	Why q?	Because r.
	3.	Why r?	Because p.
		et cetera	

Game III could obviously be carried on with any number of intermediate steps. Another kind of circle game proceeds when the opponent gives equivalents to the queried proposition each time:

IV.		White	Black
	1.	Why p?	Because $p \lor p$.
	2.	Why $p \lor p$?	Because $p \lor p \lor p$.
	3.	Why $p \lor p \lor p$?	Because $p \lor p \lor p \lor p$.
		et cetera	

V.		White	Black
	1.	Why p?	Because $\lnot\lnot p$.
	2.	Why $\lnot\lnot p$?	Because p.
	3.	Why p?	Because $\lnot\lnot p$.
		et cetera	

VI.		White	Black
	1.	Why p?	Because $\lnot\lnot p$.
	2.	Why $\lnot\lnot p$?	Because $\lnot\lnot\lnot\lnot p$.
	3.	Why $\lnot\lnot\lnot\lnot p$?	Because $\lnot\lnot\lnot\lnot\lnot\lnot p$.
		et cetera	

There may be nothing too seriously wrong with Black's circle games, depending, of course, on the relevant dialectical circumstances. But if, to win the game, Black must furnish proof of the proposition queried by White, and White must show that Black cannot furnish proof in a finite number of moves, then there could indeed be something wrong. For example, if he could not find winning proofs, Black could cheat by continuing to force White to ask the same question (or a trivially equivalent question) over and over until all the allotted moves are exhausted. White could never win any game of this sort. And Black could never lose as long as he could resort to such a manoeuvre.

But clearly the legality of such manoeuvres depends on the game one has in mind, and on how a winning strategy is defined. If one of the rules were that Black would lose if he went on filibustering without achieving some stated objective, then there might be nothing wrong with a circle game. To filibuster would be bad strategy for Black, and it might cause him to lose. But we do not *have* to forbid filibustering. Whether circle games are forbidden by the rules depends on the sort of game one has in mind, on the game's objectives, its winning strategies, and the appropriateness of other dialectical conventions.

As so far described, the fallacy of begging the question is really more like *postponing the question*. It is wrong because it allows one to stall the questioning process. But in one special kind of game, even more could be wrong with it.

Let us consider a kind of game called the *plausibility game*. White begins by asking "Why *p*?" Black must respond by furnishing a reply of the form "Because *q*", where *q* has greater plausibility than *p*, and $q \supset p$ has a (maximum) plausibility value of 10. Given White's response, the theory of plausibility tells us that the plausibility value of *p* must then be adjusted upwards to match the value of *q* (the least plausible premiss rule). The objective of this game is that Black must each time produce a premiss that is *more plausible* than the proposition he has been queried on.

What is wrong with arguing in a circle in this sort of game? Black cannot argue in a circle and play by the rules. Consider Game II (page 150). When Black offers *q* as premiss for *p*, the plausibility value of *p* must be adjusted upwards to equal the value of *q*. Thus the value of *p* cannot remain higher than the value of *q*. So when, at the next move, Black offers *p* as a premiss for *q*, he is breaking the rule that, each time, a premiss of greater plausibility must be offered. And clearly Game III, or any longer game of the same sort, will have the same problem. Arguing in a circle is more than mere filibuster: it is a direct violation of the main objective of the game, and it causes the filibusterer to lose.

Obviously Game I must be outlawed, and so must Game V. What about Games IV and VI? What is wrong with these games is not so obvious. Suffice it to say that it is difficult to see how adding V *p* or a pair of negation signs at each move is going to increase plausibility.

So in this special sort of game, the plausibility game, arguing in a circle is more than just a filibuster: it is positively wrong, by virtue of its inevitable failure to increase plausibility. One has to be careful, though, not to condemn circular arguments completely – many a dialectical game might be quite legitimate without requiring an immediate increase of plausibility at every move. Quite often, it is helpful if the rules of the game give a dialectical participant enough leeway to choose premisses that may not be very plausible, in the hope that *eventually* he may furnish a premiss of greater plausibility, and thereby vindicate all his intermediate moves.

We can appreciate why arguing in a circle is best seen as a dialectical fallacy. It is not always wrong; when it *is* wrong, it can be wrong in different ways depending on the conventions that are mutually agreed upon by the participants in a given dialectical game.

6. Evaluating Allegations of Circularity

We now know what circles look like in dialectical games, so we should be in a position to locate and identify circles in arguments. A practical method will be of some help.

In the preliminary parts of this chapter, we used the device of propositional letters – p, q, r, and so on – connected by lines to show how certain arguments can be analyzed to exhibit the structure of a *petitio*. This is a generally useful method of analyzing arguments at the practical level. Michael Scriven gives an illustration of how the method works.[13] Suppose that we are confronted with this raw argument:

> p There is hardly any product that cannot be made more cheaply by somebody who does not care about the quality.
>
> q There is no getting away from the fact that quality takes more time and costs more money.
>
> r People who believe that you can get quality for free are simply dreamers – or they've never tried to make something decent themselves.
>
> s You get what you pay for.

The first three statements seem to be premisses supporting s, the conclusion. So the structure might be graphically represented:

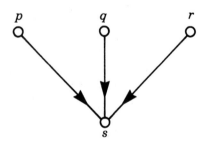

However, there is another interpretation of how the steps of reasoning might be arranged. The statement r seems to be an interim conclusion on which s, the final conclusion, is based. Accordingly, another representation of the structure of the argument is suggested:

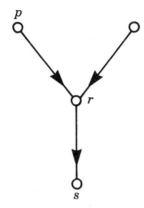

The structural diagram indicates which premisses are supposed to support which conclusion with the help of what other premisses. This example is fairly simple. But as Scriven points out, the method can be extended to longer and more complicated chains of reasoning, made up of larger units than a single claim.[14]

How is the method of argument analysis applicable to *petitio*? It should not be too difficult to see, if we look back to our first dialogue between Bob and Lester (page 134). Here, Lester's argument had the structure of a triangle.

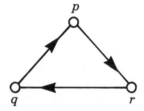

A triangle is a closed figure (as is a circle) and therefore we might suspect a *petitio*. However, if Lester had merely been arguing for *q* and *r* on the common premiss of *p*, no circle would have been involved. De Morgan arrows, which indicate the direction of premiss-dependency, would make this clear:

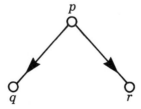

Even if Lester meant to postulate a premissory relationship between *q* and *r*, or perhaps between *r* and *q*, still no *petitio* would be decisively manifest:

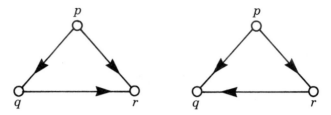

But notice that in these cases the closed nature of the argument might make us more apt to suspect that a *petitio* might be in the offing. On the other hand, we can see that the structure of Lester's argument is represented adequately by none of these diagrams. Rather, it is truly a *petitio*:

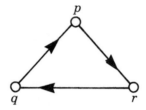

In Lester's argument, *q* is a premiss for *p*, and *r* is a premiss for *q*. But what clinches the *petitio* is that *r* is evidently meant to rest on *p* as a premiss. Here the nature of the circular dependency relationship is clearly evident.

It is fairly easy to prosecute Lester's argument on a charge of *petitio*. Once we put in the lines of premissory dependency and the arrows that indicate the direction of these relationships, the circle is closed. Combining Scriven's lines and De Morgan's arrows gives us a rather useful *graphic procedure*. In many allegations of *petitio*, however, it may be more difficult to establish whether the fallacy has occurred.

Consider, as illustrations, two argument graphs:

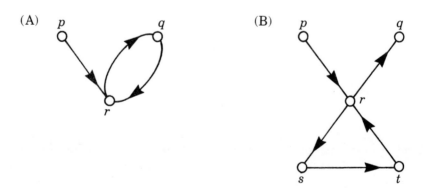

In both arguments, the arguer appears to have argued in a circle, but in each argument there is a way out of the circle. In argument (A), the arguer has proved *r* on the basis of *q*, but then he has gone on to prove *q* on the basis of *r*. So far, his argument for the conclusion, *r*, is circular. But notice that he can also prove *r* by the premiss *p* – and there is no circle in that proof. If there is a circle in an argument, *it need not be a vicious circle*. It can present no problem if a non-circular line of proof is also available. In argument (B), the arguer could have established his conclusion, *q*, by the circuitous route *p, r, s, t, r, q*. Or he could have cycled around the triangle, *r, s, t*, indefinitely, thus creating a filibuster. But again, neither the circle nor the filibuster is necessary. The arguer can avoid the cycle altogether by simply establishing *q* on the basis of *r*. And indeed, it needn't be harmful or fallacious even if he does go round the full cyclic route, *p, r, s, t, r, q*. For the cycle might be no more than a harmless detour. The main point is: even if a circle is present in an argument, the dialectical context may indicate that it is harmless.

The problem, as illustrated by our example of the dispute between De Morgan and Mill, is that the dialectical background of an argument may very often be unclear to the participants or critics. By means of a retroductive analysis of the premisses, we can attempt to determine the dialectical background that the proponent of the argument is relying on. What possible basis could he have derived his premisses from? After this dialectical procedure has been carried through, we can establish whether the chain of arguments constitutes a *thema* that commits a *petitio*. Thus it is one thing to suspect that an argument might commit a *petitio* and to challenge its proponent by raising the question. It is quite another to press a charge of *petitio* by successfully closing the circle of actual steps of argument carried out by that proponent. To achieve such success, the dialectical technique of Socratic questioning may often be essential.

Exercise

1. Many a logic text offers the following sequence as an illustration of the *petitio principii*. When asked by his bank manager for a credit reference, Smith replies, "My friend Jones will vouch for me." Manager: "But how do I know I can rely on a reference from Mr. Jones?" Smith: "Oh, no problem – I can vouch for Mr. Jones myself!" Evaluate whether Smith has committed a *petitio* by supplying several alternative dialectical contexts for the argument.

2. Black and White are playing a plausibility game, and the initial plausibility values are as follows: $p = 8$, $q = 7$, $r = 6$, $s = 5$. Every conditional has a plausibility value of 10. Here is a sequence of the game:

White	Black
1. Why s?	Because r and $r \supset s$.
2. Why r?	Because q and $q \supset r$.
3. Why q?	Because p and $p \supset q$.
4. Why p?	Because s and $s \supset p$.

Is this a circle game? Draw a graphic diagram to clarify the relationships among p, q, r and s. If it is a circle game, is there anything wrong with it?

3. Black and White are playing a plausibility game, and the initial plausibility values are as follows: $p = 8$, $q = 5$, $r = 7$, $s = 6$. Every conditional has a plausibility value of 10. Here are three sequences of the game:

I.

White	Black
1. Why q?	Because s and $s \supset q$.
2. Why s?	Because r and $r \supset s$.
3. Why r?	Because q and $q \supset r$.
4. Why q?	Because p and $p \supset q$.

II.

White	Black
1. Why q?	Because s and $s \supset q$.
2. Why s?	Because r and $r \supset s$.
3. Why q?	Because p and $p \supset q$.
4. Why r?	Because q and $q \supset r$.
5. Why s?	Because r and $r \supset s$.

III.

White	Black
1. Why q?	Because s and $s \supset q$.
2. Why s?	Because r and $r \supset s$.
3. Why r?	Because q and $q \supset r$.

For each sequence, draw a diagram to represent the flow of argument. Then evaluate whether there is a circle. If there is, decide whether it represents a fallacious manoeuvre.

Notes

1 It is also called begging the question, circular reasoning, *circulus probandi*.
2 Iving M. Copi, *Introduction to Logic*, fifth edition, New York and London, Macmillan Publishing Co., Inc. and Collier Macmillan Publishers, 1978. By permission of the author.

3 "Anthropomorphism", from the Greek *anthropomorphos* meaning of "human form". Thus the attribution to God of human characteristics.

4 Augustus De Morgan, *Formal Logic* (London: Taylor and Walton, 1847), page 254.

5 Alfred Sidgwick, *Fallacies* (New York: D. Appleton & Co., 1884), page 194.

6 Morris Cohen and Ernest Nagel, *An Introduction to Logic and Scientific Method* (New York, Harcourt, Brace & World, 1934), p. 177.

7 Other examples can be found in the writings of Sextus Empiricus.

8 De Morgan, page 254.

9 De Morgan, page 257.

10 See William and Martha Kneale, *The Development of Logic* (Oxford: Clarendon Press, 1962), p. 163.

11 Aristotle, *Posterior Analytics*.

12 P.T. Geach, *Reason and Argument* (Oxford: Blackwell, 1976), Chapter 14.

13 Michael Scriven, *Reasoning* (New York: McGraw-Hill, 1976), page 157ff.

14 See Scriven, see also R.H. Johnson and J.A. Blair, *Logical Self-Defense* (Toronto: McGraw-Hill Ryerson, 1977), page 176ff.

Chapter Eight
Formal and Informal Logic

At this point in our discussion a certain question should be clarified. How formal should the theory of fallacies be? As the need to develop precise guidelines for the models of argument involved in the fallacies becomes greater, we tend to move in the direction of "formal" logic, a logic in which precise rules and conventions are indicated. But many textbooks on the fallacies resist this direction of development; they adhere to the tradition that thinks of the fallacies *informally*, and considers them outside the domain of formal treatment.

In order to resolve this apparent conflict, we need to know what, exactly, is meant by the terms "formal" and "informal" as applied to the study of logic. First, we will explain what a formal logic is and what is meant by "logical form". Some fallacies have to do with the meanings of words or sentences in a natural language, like English. We look at such fallacies to see whether they are amenable to treatment by formal methods. We will see that formality admits of degrees, that the logic of the fallacies should be formal to some extent, but that its degree of formality is determined by various factors. Formality cannot be adhered to as strictly as some philosophers might like, but it cannot be abandoned, either. So the logic of the fallacies will be partly formal.

As the chapter evolves, three concepts will be explained:

1. Logic as a *theory*
2. Logic as a *formal* theory
3. The logic of the fallacies as *informal* logic

Logic is a science of reasoning. As a science it has a general, rather than a particular, orientation. Physics, for example, tells you about mass and energy *as such*, and not just about this or that chunk of granite or this or that beam of sunlight. Similarly, the logic of deduction attempts to reveal the most universal truths about deduction. It does this by studying the forms of arguments.

We shall begin very simply with propositional logic.

158

1. Logical Form

Here is a very simple deduction, one we have already met with.

> (D1) All the change in my pocket is dimes.
> Therefore, all the change in my pocket is dimes.

Here is another:

> (D2) All blondes are curvy
> Therefore, all blondes are curvy

Another still:

> (D3) It is sexist to say that all blondes are curvy
> Therefore, it is sexist to say that all blondes are curvy

All good deductions, to be sure.[1] But are they trivial, as well? There is indeed a kind of triviality involved here. Not only are these very easy and obvious deductions (though they certainly are, in that sense, trivial); not only do the conclusions merely repeat the premises (though in that sense, too, they are trivial). There is another kind of triviality here. Suppose we started out to teach the mysteries of deduction by telling you that (D1) is correct; suppose we then ask you to consider (D2), and assured you that it is also correct; suppose our next step was to ask you to examine (D3), and that we insisted that it, too, is correct. Your likely response would be, "For heaven's sake, teach us something *new!*"

And you would be right. Though (D1), (D2) and (D3) are *different deductions*, they all share a *common pattern*:

$$p$$
$$\therefore \overline{p}$$

It is the *structure* of this common pattern that accounts for the correctness of our three deductions; that is, our three deductions are correct *because* they share this pattern.

Logicians have a special term for a pattern of argument that makes any conforming argument a deductively correct argument. Such a pattern is called the argument's *logical form*. It is quite important to be clear about logical form. Not every common pattern is of interest to logicians; indeed, a pattern is interesting only when it *makes an argument logically interesting*. Here is a pattern for an argument:

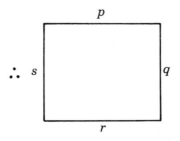

This pattern is not logically interesting; in fact, it is logically incoherent. The pattern p, $\therefore$ p is logically interesting. *Any* argument that follows this pattern is a *correct deduction*.

Thus (D1), (D2), and (D3) are, *by their logical form*, deductively correct arguments; so is *any other* argument of that same form.

Here are three more deductively correct arguments.

(D4) Everybody loves my baby and my baby loves me only.
$\therefore$ My baby loves me only.
(D5) The cat is on the mat and the dog is in the manger.
$\therefore$ The dog is in the manger.
(D6) Sue punched Bill and Hank jumped up and down with glee.
$\therefore$ Hank jumped up and down with glee.

Correct deductions all, by virtue of this logical form:

p and q

$\overline{\phantom{p \text{ and } q}}$

$\therefore q$

By considering the *form* of an argument, we can discuss general *patterns* of argument, and thereby study formal logic in a scientific way. For, as we saw in the case of deductive logic, once we have translated an argument from English (or Hindu, or whatever) into a mathematical language, we can test the form of the argument for correctness. And any argument that has a correct form must be deductively correct.

Arguments in natural language are often hard to interpret or formulate clearly; they often need a lot of "cleaning up", as we have seen. But once we have done our translations, we can deal with the pure form of the argument. Assigning it to the categories of *correct* or *incorrect* can sometimes be accomplished with the help of exact methods.

Here are some examples. Consider:

(D7) Sue went to the movies or Sue left town.
It is not the case that Sue went to the movies.
Therefore, Sue left town.
(D8) Betty is a genius or Hank is lucky.
It is not the case that Betty is a genius.
Therefore, Hank is lucky.
(D9) Tay-Sachs disease is infectious or spina bifida is congenital.
It is not the case that Tay-Sachs disease is infectious.
Therefore, spina bifida is congenital.

These deductions, too, are quite correct. The logical form that makes them correct is called the *disjunctive syllogism*.

p or q
It is not the case that p

$\overline{\phantom{\text{It is not the case that } p}}$

$\therefore q$

Look again at example (D9). This example was chosen deliberately, in the belief that most general readers would not know what Tay-Sachs disease or spina bifida are. After all, these are technical medical terms. Still, it is a fair bet that even those who did not understand what these sentences mean had no difficulty whatever in seeing that the deduction is correct. This is an extremely important point, well worth repeating.

Some deductions can be determined to be correct independently of the meaning of their constituent sentences.

But now consider this "argument".

(D9a)   Tay-Sachs disease is infectious *glarg* spina bifida is congenital.
Ooglah Tay-Sachs disease is infectious.
Therefore, spina bifida is congenital.

Here is an "argument" that very few readers – in fact, not one – will fully understand. And no one will be able to determine whether it amounts to a correct deduction. Even if you had known the meaning of "Tay-Sachs disease" or "spina bifida", you still would not have been able to tell whether you had a correct deduction.

Whether an argument is a correct deduction does not always depend on the meaning of all its constituent words and expressions. But it *does* depend on the meaning of *some* of them.[2]

Let us look again at our disjunctive syllogism:

p or q
It is not the case that p

$\therefore q$

Notice that the letters "p" and "q" are just letters, and *do not mean anything*. Notice, as well, that the word "or" and the phrase "it is not the case that" *do* have meaning. In fact, it is the meaning of these words that assures us that the disjunctive syllogism is a deductively correct logical form.

Let us compare Figure Three and argument (D9). Notice that the ps and qs of Figure Three correspond to the elements in (D9) whose *meanings* are *not* necessary for deductive correctness. Thus those elements in (D9) whose meanings do not count are replaced, in Figure Three, with other meaningless elements. Our comparison of Figure Three with (D9) emphasizes our original point: the meanings of the phrases "Tay-Sachs disease is infectious" and "Spina bifida is congenital" do not affect the deductive correctness of the argument.

It is also important to notice that precisely those elements of (D9) that do not affect deductive correctness have been changed in both (D7) and

(D8). The three arguments vary among one another precisely where their meanings do not count.

Logicians call the letters in a logical form such as Figure Three sentence variables.

Let us now examine the respects in which (D7) to (D9) and the disjunctive syllogism resemble one another and the respects in which they do not vary (in other words, respects in which they hold certain features *constant*). Simple inspection tells you that they share, in the same places, occurrences of the word "or" and the prefix "not". Since these are constant elements running through (D7) to (D9) and the disjunctive syllogism, logicians call them *sentence constants*.

Look again at any two of the arguments that we have been discussing, (D7) to (D9). (D7) is the deduction:

Sue went to the movies *or* Sue went to town.
Sue did *not* go to the movies.
Therefore, Sue went to town.

It is easy to see that the argument (D9) *could be constructed from* (D7) by a very simple operation, namely, by *replacing* the sentences "Sue went to the movies" and "Sue went to town" by the sentences, "Tay-Sachs disease is contagious", and "Spina bifida is congenital." Thus:

Tay-Sachs disease is contagious *or* spina bifida is congenital.
Tay-Sachs disease is *not* contagious.
Therefore, spina bifida is congenital.

There are two points about this operation of replacement to pay special attention to:

First, the constants "or" and "not" are *not* touched; they are left alone.

Second, the replacement is *uniform*. That is, any sentence replacing a given sentence in (D7) replaces every other occurrence of it in (D7). This is what is meant by *uniform replacement*.

We can now see that *any one* of the arguments (D7) to (D9) could have been attained from any of the others by the operation of uniform replacement of their *non-constant* elements.

Similarly any of those arguments could have been attained from the logical form (disjunctive syllogism) by uniform replacement of *variables* with *sentences*. These arguments would then be called *substitution instances* of their respective logical forms.

What, then, is logical form?

Logical form is a configuration of *variables* and *constants* such that *uniform replacement* preserves deductive correctness. That is, if a logical form is deductively correct, so, too, is any argument attained from it by uniform replacement of its variables with sentences.

Similarly: if a given logical form is deductively correct, so, too, is any other *logical form* attained from it by the uniform replacement of variables with variables.

It is now easy to understand why the logical form of a good deduction is interesting to a logician – it is *because* of the form that the argument is deductively correct. And the *reason* logical form *makes* deductive correctness is that the form itself is deductively correct, and uniform replacement preserves deductive correctness.

Exercise

Which of the argument forms (1) and (2) is a substitution instance of which argument form (a) to (f) below?

(1) If Harry has gone, Zelda will leave.
Zelda will not leave.
Therefore, Harry has not gone.

(2) Either Harry has gone or Zelda will leave.
Harry has not gone.
Therefore Zelda will leave.

(a) If p then q
p
$$\frac{}{\therefore q}$$

(b) p and q
p
$$\frac{}{\therefore q}$$

(c) p
p
$$\frac{}{\therefore p}$$

$$(d) \quad p \text{ or } q$$
$$\text{Not-}p$$

$$\therefore q$$

$$(e) \quad q \text{ or } p$$
$$\text{Not-}q$$

$$\therefore p$$

$$(f) \quad \text{If } p \text{ then } q$$
$$q$$

$$\therefore p$$

2. Formal Logic

We use the theory of deduction to explore arguments from the point of view of their formal attributes. We may say, then, that the theory of deduction is a *formal theory*, and that deductive sentence logic is (a branch of) *formal logic*.

Here is a rough *rule of thumb:*

An argument admits of study from a formal point of view to the extent to which it can be checked for logically interesting properties without attending to the meaning of its constituent elements.

The rule of thumb at once provides that the arguments (D1) to (D3) admit of *completely* formal treatment, for they all have the same form:

$$p$$
$$\therefore p$$

Hence, none of the three arguments depends for its deductive correctness on the meaning of any constituent element.

Formality is a matter of degree. The arguments (D4) to (D6) could all be tested for deductive correctness; we could ignore the meaning of all constituent elements *except* the single constant, "and". And arguments (D7) to (D9) also admit of a perfectly adequate test of correctness, independent of the meaning of all constituent elements *except* the two constants, "or" and "not".

Consequently, all of these groups of arguments admit of *formal* treatment, but in *different degrees*. There is another way of phrasing our rule of thumb:

> The formality of an argument is indirectly proportional to its dependency on its constants for such properties as deductive correctness.

Let us now look at an argument that, while a good deduction, seems to admit of relatively little formal treatment.

> The shirt is red.
> Therefore, the shirt is coloured.[3]

As we said, a good deduction. But no formal logician would be much interested in it. His decided preference is for highly formal arguments. But, we may ask, what *justifies* such favouritism? Quite simply, formal arguments admit of assessments that are *very general*, and generality is one of the hallmarks of true science. Logic, like any other science, aspires to the deepest and most general truths of its domain, and logical form is a logician's method of attaining such general truths. A couple of examples will make the point clear.

Let us look again at example (D1).

> All the change in my pocket is dimes.
> Therefore, all the change in my pocket is dimes.

This is a correct deduction: it is correct by virtue of the logical form "*p*, therefore *p*". Now, because deductive correctness is preserved under uniform replacement of non-constant elements, we can *also* say that every deduction of that same form is likewise correct. This is a *law* of the theory of deduction.

> **LAW**
>
> All arguments in which the conclusions are identical to the premisses are correct deductions.

Now look at our second argument:

> Everybody loves my baby and my baby loves me only.
> Therefore, my baby loves me only.

This, too, is a correct deduction. It owes its correctness to the logical form, "*p* and *q*; therefore *q*", by which we are introduced to another law of deduction:

> **LAW**
>
> All arguments in which the premiss is composed of one sentence followed by the constant "and" and followed, in turn, by a sentence, and in which the conclusion is that second-mentioned sentence, are correct deductions.

3. Interim Summary

(1)   A deduction is correct by virtue of its logical form.

(2)   A logical form is a configuration of sentence variables and sentence constants, in which uniform replacement of sentence variables with either sentences or (other) sentence variables preserves deductive correctness.

(3)   An argument admits of formal assessment if the determination in question can be made by appealing to the argument's logical form.

(4)   A deduction admits of formal treatment (or, more briefly, *is* a "formal argument") to the degree that its correctness does not depend on the meanings of its constituent elements. Consequently, a rough rule is this: the greater the proportion of variables to constants in an argument's logical form, the greater is the degree to which that argument is a formal argument.

Moral of Point Four: It is now clear that the logic of deduction need not deal directly with actual, everyday arguments. It suffices that the logic of deduction deal directly with logical form. For once the deductively correct logical forms are known, it is also known which *arguments* are deductively correct. Bearing in mind that uniform replacement preserves deductive correctness, any argument is deductively correct if it arises from the uniform replacement of variables with sentences from a deductively correct logical form.

---

(5)   There is a precise and specific respect in which deductive logic is formal: *deductive logic deals mainly, and in the first instance, with logical form.*

(6)   To discover the deductive correctness of a given logical form is also to discover that *any and all* arguments arising from that form by the uniform replacement of variables by sentences are likewise deductively correct arguments. Thus the formality of formal logic makes possible the discovery and articulations of *very general truths*, and this means that logic has a scientific basis and a law-like orientation.

Formal logic removes all possibility of ambiguity by fixing the meanings of the logical constants of a formal theory so that the correctness or incorrectness of an argument expressed in that theory does not depend on anything that might change. Correctness is determined exclusively by the form of the argument.

There is a problem with fallacies, however. Their medium is usually a natural language such as English, where there do not seem to be precisely defined logical constants. Can formal logic be applicable to such arguments? Let us test the question by turning to a fallacy that depends expressly on ambiguities of natural language.

4. Equivocation

The fallacy of equivocation may occur when a word that can be interpreted or defined in more than one way plays a critical rôle in an argument. A word like "bank" is ambiguous – it may be taken to mean "bank of a river" or "savings bank", depending on the context of its use. Ambiguity may be harmless enough, provided it is clear what is meant, or provided that ambiguity doesn't matter. Sometimes, however, the meaning is not clear, and ambiguity does matter. Take this argument:

> (1) The end of a thing is its perfection.
> Death is the end of life.
> Therefore, death is the perfection of life.

Something seems badly wrong with this argument; it is deductively correct, and the premisses seem true enough, yet the conclusion is outrageous, and seems to be false or wrong. What is wrong is not hard to see. The word "end", in the first premiss, means *goal* or *purpose*, whereas the word "end" in the second premiss means *termination* or *last point*. The ambiguity of a word – "end" – is not harmless at all. Rather it is a pernicious confusion. What seems to be a correct and plausible argument (if you do not inspect it too closely) is in reality a tricky deception.

But what is really wrong with the argument, if it is deductively correct and both premisses are plausible? It is not an argument at all! It is not *one* argument, but *four* arguments. This multiplicity arises from the fact that the word "end" is ambiguous and thereby allows four interpretations of what seemed to be one argument. Here are the other four arguments:

> (2) The goal of a thing is its perfection.
> Death is the goal of life.
> Therefore, death is the perfection of life.
> (3) The termination of a thing is its perfection.
> Death is the termination of life.
> Therefore, death is the perfection of life.
> (4) The goal of a thing is its perfection.
> Death is the termination of life.
> Therefore, death is the perfection of life.
> (5) The termination of a thing is its perfection.
> Death is the goal of life.
> Therefore, death is the perfection of life.

Taken individually, none of these arguments is any good. Argument (2) has an implausible premiss, "Death is the goal of life". Argument (3) also has an implausible premiss, "The termination of a thing is its perfection". Like Arguments (2) and (3), Argument (5) also has an implausible premiss, "Death is the goal of life". Only Argument (4) has two quite plausible premisses. But Argument (4), like Argument (5), is not deductively correct. Only Arguments (2) and (3) are deductively correct. No matter which argument you choose, you can't find one that is deductively correct and that also has plausible premises.

No matter which way the person who is confronted with this argument turns, he is confronted with an unsatisfactory argument. On one side, he is tugged towards Argument (4) because of its plausible premises. On the other side, he is tugged towards Arguments (2) or (3), because of their deductive correctness. The urge to give in to the fallacy is to try to have both plausible premises (4) and deductive correctness (2 and 3) by rolling up more than one argument into (1). It seems that in Argument (1), we can have plausible premises and a correct argument. Equivocation is the easy way out.

In our example, the equivocation can be traced to multiple meaning in one word, "end", which occurs in the argument. A similar, and rather delightful, equivocation turns on ambiguous use of the word "cross": "A person undertakes to cross a bridge in an incredibly short time: and redeems his pledge by crossing the bridge as one would cross a street, that is, by traversing the breadth."[4] Here the word "cross", as in "to cross a bridge", has a special meaning that could not be precisely inferred from its individual dictionary meaning. The context of the sentence as a whole that it occurs in leads us to interpret it as plausibly meaning "cross" in the sense of going from one *foot* of the bridge to the other.

Sometimes it is not one word, but the whole sentence, that is ambiguous. For example, consider this argument:

> Everything has a cause.
> Therefore, there is something that causes everything.

The premiss may appear plausible enough as a principle of scientific explanation, or perhaps as an article of scientific faith. It says that any phenomenon, if you examine the phenomenon carefully enough and learn enough about it, the premiss says, you can link that phenomenon to some causal antecedent that might help to explain how it happened. But even if this proposition is true, the conclusion – that there is some *one* thing that causes everything — need not be true. The premiss says that every event, E_i, has a cause, C_i (where "i" means any number).

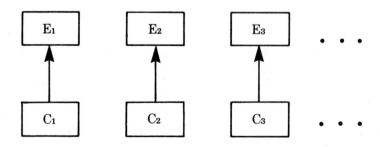

But the conclusion says that every event has *the same* cause:

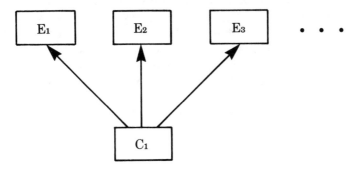

If the argument were correct, it could prove the existence of God. But, unfortunately for natural theology, the argument is an equivocation.[5] The equivocation here is one that turns on an ambiguity in the structure of the entire proposition, not on just one word. The premiss says: "For [anything you like], something causes it". But the conclusion says: "There is something that causes [anything you like]".

5. Amphiboly

Traditionally, logicians have distinguished between equivocation and *amphiboly*. The fallacy called amphiboly occurs where a double argument turns on grammatical rather than semantic ambiguity. For example, the slogan "Save soap and waste paper" is said to be *grammatically* ambiguous. The slogan can be interpreted in several different ways, depending on whether "waste" is interpreted as a verb or as an adjective. But the distinction between grammar and semantics is not clear: is the slogan's ambiguity grammatical or semantical (meaning-based)? Traditional logic texts have never been able to furnish any very convincing or realistic examples of amphiboly.

Many of the examples of amphiboly cited by textbooks are, in fact, not arguments at all; they are, instead, ambiguous statements. Some examples:

> LAUNDROMAT SIGN: Customers are required to remove their clothes when the machine stops.
> ROADHOUSE SIGN: Clean and decent dancing every night except Sunday.

These sentences are not fallacies – because they are not arguments. They are *grammatically* ambiguous sentences, and they could be misleading. But that does not mean they are incorrect arguments; nor does it mean that amphiboly represents a kind of fallacious argument. Amphiboly, like the *ad populum* and the *baculum*, is a case of an alleged fallacy that may not be a fallacy at all.

Some texts do give examples of amphiboly that are stated as arguments. For example, one text cites this argument: "Thrifty people

save old rags and waste paper; therefore, thrifty people waste paper."[6]
The argument has the form "$p \land q$, therefore q", a deductively correct
form of argument.

But we certainly would be making a mistake if we accepted the
conclusion on the basis of *one* interpretation of the premiss. The case of
amphiboly is therefore similar to equivocation – there is more than one
argument here, and mischief arises when we lump together two or more
arguments, none of which is good in all respects. Thus amphiboly is
similar to equivocation. But the ambiguity is grammar-based rather
than meaning-based.

Grammatical factors do sometimes play a rôle in equivocation.
Consider this argument:

> Schemers are not to be trusted.
> Bob has devised a scheme to prevent his vegetables from rotting.
> Therefore, Bob is not to be trusted.[7]

The arguer has assumed that everyone who devises a scheme is a
schemer, where "schemer" is defined in a way that would make the first
premiss plausible. Some would call this argument an instance of
amphiboly, but we prefer to think of it simply as an equivocation with a
grammatical twist.

To sum up, then: *equivocation and amphiboly are fallacies that occur
when we take as one argument what is, in reality, more than one
argument*. Equivocation and amphiboly are fallacies of meaning and
grammar, which arise through ambiguity or multiple meaning.
Sometimes what is taken to be an argument is really no argument at all;
we do not have *enough* argument. But in equivocation and amphiboly,
we have *too many* arguments at once. Before evaluating any argument,
we must always be sure to locate and identify the argument.

Exercise

Identify and explain the fallacies in the following arguments. List
as many ambiguities as you can find.
1. The press should print news that it is in the public interest to
   be disseminated and discussed. The public interest in Margaret Trudeau is intense. Therefore, the press should print
   news on Margaret Trudeau's activities.
2. Saints and heroes who have sacrificed everything for others
   did so because they wanted to do so. They were really acting
   selfishly. Consequently, no free human action is unselfish.
   Altruism and human decency do not exist!
3. Jones shows discrimination in his taste in clothes. Do you
   think he could be guilty of discrimination as a violation of
   human rights in his business activities? Never trust a man
   who dresses too well!

4. For sale: Newfoundland dog. Will eat anything; very fond of children.
5. Hugo was born in Victoria Hospital, then in St. Vital. Therefore, Hugo was born twice.
6. My uncle said, "The shooting of the hunters is terrible." If he believes that the shooting of the hunters is terrible, he must be against the killing of wildlife. Therefore, my uncle is an animal lover.
7. All men are not feminists. Therefore, no men are feminists.

Equivocation and amphiboly are interesting to the moral philosopher because they represent ways of not quite telling the truth, yet they may fall just short of lying. Here are two interesting mediaeval examples:

(1) Knowing the answer is yes, one nevertheless answers a question by saying: "I say no."
(2) A monk is hiding a fugitive; to the query of whether the fugitive is there, the monk replies, *"Non est hic."*

The straightforward meaning of the Latin phrase is "He is not here," but the monk consoles himself with the thought that his words could be taken to mean "He is not eating here." Most of us would probably say that the monk had lied, but it depends on what you mean by *lying*.[8]

Such arguments leave the equivocator (or amphibolator) a handy excuse – he or she can always say, "Oh, but I meant you to take it the other way. How unfortunate you took my phrase in a way I did not intend." When King Croesus asked the Delphic Oracle whether to wage war against the Persians, the clever Oracle gave the following advice: "If Croesus should wage war against the Persians, he would destroy a mighty empire." Thus the Oracle had an out in the event that Croesus waged war and lost. The Oracle could always claim that the "mighty empire" was that of Croesus.

6. Why Equivocation Is a Problem for Logic

The examples of equivocation and amphiboly we have studied so far are simplistic "text-book" cases, which would be unlikely to deceive an attentive, perceptive, and critical audience. Really interesting cases might be expected to occur in contentious and unclear argumentation, where the terms are hard to define or grasp, and where premises are complex enough for shifts of meaning to pass unnoticed. In these non-trivial cases, it may be difficult or even impossible to pin down, prosecute, or refute any charge of equivocation, for the meanings of the key words or propositions may themselves be a subject of controversy.

If all the terms of a dispute are clearly defined at the outset, and all participants agree precisely on the definitions, then the matter of equivocation can be sorted out. But how often, in an argument, do the participants begin with that degree of terminological precision and agreement? Almost never, unfortunately!

Even worse, sometimes arguers change the meaning assigned to a term in mid-argument, or introduce a *stipulated* meaning. So the practical problems of evaluating an allegation of equivocation are not to be treated lightly. Certain dialectical conventions concerning the meanings of terms used in an argument will have to be clearly fixed and stated at the outset of an argument if equivocation is to be dealt with.

Couldn't the participants in an argument agree to use some natural language, say English, and adhere to the linguistic conventions of English? There is a problem with this proposal: it leaves us still confronted with the problem we began with, namely the problem of knowing exactly what the agreed-upon conventions of English are. *Because* controversial terms in English are ambiguous, unclear, and disputable, we get into trouble with equivocations.

Someone else might suggest utilizing a highly precise language where *all* the meanings of the constants are precisely defined – say, classical deductive logic. Such precision would truly allow us to eliminate any problems we might have with equivocation. But we saw that classical deductive logic has some severe limitations as an exclusive method of dealing with arguments. The basic limitations, we remember, are two. First, we had to translate from natural language into the language of classical logic, and this meant that many aspects of natural language were simply left by the wayside. Second, classical logic is exclusively concerned with truth-values; as a result, it is always open to the fallacy of irrelevance *(ignoratio elenchi)*.

But there is a way to solve the problem. Deductive logic, inductive logic, and plausibility theory must be absorbed into dialectic; thus dialectic can take advantage of the precision of such methods, and still maintain its generality and applicability to arguments in their natural setting of contentious disputation.

The ultimate goal of "logical analysis" may be translation into a language with an exactly specified structure in which well-defined procedures, with constants and variables specified in advance, regiment if not totally eliminate ambiguity. In practice, however, argumentation takes place in a less rarified atmosphere, and our guidelines must be applicable to practice. Characteristically, argument in natural language can tolerate ambiguity (and hence equivocation). But we will now show that, even in practical logic, if we are to have what may be considered an argument in a natural language, certain boundaries must be set on what that language may be said to comprise.

It is now useful to have at hand the distinction between *object-language* and *metalanguage*. For present purposes, it suffices to say that any of the *object-languages* we might be interested in are languages in which are produced the arguments that we wish to decide or evaluate. And the language in which such descriptions and evaluations are conducted is called a *metalanguage* relative to that object-language. Typically, of course, a person's object-language and metalanguage are one and the same, for example, English, but this need not be so.

- The logician Alfred Tarski has shown that we must distinguish between expressions of the *object-language* and expressions of the *metalanguage* if we are to avoid the possibility of inconsistency within the language. Hence any language suitable for argument should not be viewed as semantically closed. What Tarski meant by a semantically closed language is one in which the semantic conventions governing the meaning and use of the terms are formulated in that very language itself. Or, to put the point another way: a language is semantically closed if it is its own metalanguage. If we want to investigate the semantics of a language, and if we use that language itself as the vehicle of semantic investigation, we will be caught in a kind of vicious circle. Tarski showed the results of such semantic self-closure in an example, the *Paradox of the Liar*.

(1) Every declarative sentence of English is either true or not true, but not both.
(2) Moreover, what you see written in the rectangle on this page is a declarative sentence of English.

> The sentence within the rectangle on page 173 of *Argument: The Logic of the Fallacies* is not true.

Let us call the sentence in the rectangle "E". Assertions (1) and (2) clearly imply that

(3) Either E is true or E is not true.
(4) If E is true, then what it says is true, namely, that E is *not* true.
(5) But if E is not true, then things are as E says they are; thus, E *is* true.

Thus, if E is true then E is not true, and if E is not true then E is true. And E is either true or not true, but not both. Consequently,

(6) E is true and E is not true.

But this is a *contradiction*.[9] What is wrong in the liar paradox? The semantic conventions that govern which sentences are to be counted as true are formulated in the *very same language* in which certain true statements are made; that is, the language in which the paradox arises is semantically closed. But the paradox shows that *any* language that is semantically closed contains a logical contradiction. Therefore, such closure must not be allowed.

Exercise

Can you say more precisely where the deduction went wrong in the liar paradox, and what factors make it a paradox?

The lesson is that dialectic must proceed by presupposing some clear semantical conventions. And these conventions may need to be stated in a language different from the actual language used for dialectical interchanges. In particular, certain restrictions must be laid down to prevent linguistic anarchy or the attendant logical contradictions and fallacies of semantic closure. Here are some restrictions.

(I)   The object-language of a disputation must be at least roughly specified. That is, the participants must agree that the argument is to take place within some natural language (English, Swahili); a natural language enriched by a well-defined technical vocabulary (English plus NASA jargon); or some quasi-artificial language, in which grammar and semantics must be at least roughly defined. Even a purely stipulative language might be allowed, if it is clearly stated what that language is.

(II) It must be possible to specify the grammar and semantics of the language at least to the extent that a claim of ambiguity can be supported or arguably rejected. That is, there must be some *data* against which claims of ambiguity can be adjudicated. There must be some form of data-processing procedure or some set of agreements to handle questions of ambiguity independently of the mere say-so of the disputants.

(III) On the basis of (II), certain common ambiguities that could occur in the disputation might be classified and recognized at the outset as ambiguous by participants in the argument.

(IV) In cases of argument within stipulated or artificial languages, the structure of the language should approximate a specified structure of the kind currently familiar in axiomatic systems as far as possible. *Ad hoc* stipulations must be banned by minimal constraints on the introduction of new or unfamiliar terms.

(V)   A stipulated meaning must be marked as such, and appropriate rules and definitions introduced by the proponent of the argument. Some metalinguistic conventions concerning the setting out of definitions, axioms, and the like, should be available. We are thinking here of procedural rules of dialectic of the type discussed in our games of formal dialectic. For some purposes, these rules need not be formalized very strictly.

(VI) The analysis and discovery of equivocation proceeds by being clear about what is *stated* by the premisses and conclusion of an argument. As a beginning, therefore, premisses and conclusion should be isolated into distinct statements. Very often what is put forward as an "argument" is really not an argument at all, but rather a command, threat, promise, or some form of utterance that does not express a statement. To recognize equivocation, we need to clearly appreciate, and be able to apply, the distinction between *truth of a statement* and *correctness* (validity) *of an argument*. Equivocation thrives when this distinction is blurred in practise.

(VII) Metaphorical use of terms and the use of peripheral meanings, idioms, and analogies in argumentation need to be looked at with caution. In such contexts, equivocation should be looked for with special care.

(VIII) Grammatical peculiarities often contribute to serious equivocations (as in the "schemers" example).

Equivocation may seem quite simple and theoretically untroublesome on the surface – it could simply be described as the appearance of validity conjured up through the conflation of what are, in reality, two or more arguments. Yet in practice, to avoid equivocation in argumentation is far from simple. Nor are guidelines yielded by this simple conception for establishing when equivocation has actually occurred in a real specimen of argumentation. We have offered a set of working guidelines, but obviously the guidelines are barely the beginning. In fact, successful prosecution of the *charge* of equivocation is often very difficult to carry out against a determined opponent.

Thus a dialectic that can cope with equivocation must be formal, to the extent that guidelines and agreed-upon conventions are stated precisely. An object language must be clearly demarcated by means of another level of language (the metalanguage), which should be thought of as distinct from the object language itself. The level of formality achieved in dialectic will not be that of formal deductive logic. Yet, insofar as formal deductive logic can be used in dialectic, a level of generality and formal precision of meaning will be attained, which can eliminate equivocation altogether.

7. *Ignoratio Elenchi* Again

We have seen that dialectic can cope with equivocation only by stating its rules and conventions *precisely*. Dialectic stands to gain, therefore, by incorporating the best known body of clearly formulated semantic structures appropriate to argument – namely, deductive logic. By collaborating, dialectic and deductive logic can each strengthen the weaknesses or limitations of the other. But what about the notorious fallacy of *ignoratio elenchi*? Can dialectic cope with that?

We remember that *ignoratio elenchi*, like equivocation, is a fallacy of meaning. But in the *ignoratio elenchi*, the fault is that the premises are irrelevant to – that is, too widely separated in meaning from – the conclusion that is supposed to be proved.

Dialectically, an arguer who uses an *ignoratio elenchi* seems to presuppose that any argument begins with a *topic*, or set of topics, which represent *what the argument is about*. When one participant starts to stray too far away from this given set of topics, he may fairly be accused of *ignoratio elenchi*. Take the example we studied in Chapter III, where Senator Paul Martin responded to the suggestion that Windsor is an ugly and grimy city by first mentioning the flower park in Windsor, and then by speaking of the town's fine schools and hard-working, tolerant citizens. There was a charge of *ignoratio elenchi*: he had strayed off-topic in attempting to bring in these other (admittedly worthy) qualities of Windsor, after he had made his point about the park. The fallacy, then, amounts to this: the topic of the argument is Windsor, its beauty or ugliness. But not every aspect of Windsor is included in this topic. The virtues of tolerance, diligence, and educational attainments are not included in this topic. Therefore, a reference to them is irrelevant. Similarly, if Senator Martin includes in his reply that Peterborough is a gorgeous specimen of urban beauty, he has strayed off-topic. True, he includes beauty as a topic of his argument, but he is off the mark in bringing in Peterborough when the topic is Windsor.

Clearly the question of *how far an arguer is off-topic* can be a matter of degree. Suppose Senator Martin had added to his argument the statement, "And then, too, at lower than usual temperatures, the human brain may be deprived of an oxygen supply for relatively prolonged periods without the onset of brain death." There seems to be no connection at all between his statement and the issue of the beauty of Windsor. The failure of relevance is extreme.

In order to control and evaluate the *ignoratio elenchi*, dialectic must specify not only conventions concerning the language in which argument is to take place, but also a set of topics within that language. These topics must be defined as clearly as possible by the participants. If a set of premises shares *no* topic in common with a conclusion, we can definitely say an argument is an *ignoratio elenchi*. If the premises share some topics but are divergent in other topics, other factors may have to be taken into account to settle the issue of relevance.

Another major factor in determining relevance is the correctness of the argument. Sometimes, even though there is topic overlap, we feel that the argument misses the point because it fails to be a *correct* argument. In our example, the statement "Windsor has hard-working citizens" does share a topic (Windsor) with the conclusion to be argued for, "Windsor is beautiful." But the problem is with the argument:

> Windsor has hard-working citizens.
> Therefore, Windsor is a beautiful city.

The argument is not deductively correct. The plain fact is that this is a dreadful argument.

Mere subject-matter overlap (or relatedness) between premises and conclusion does not constitute a sufficient condition of correct argument. Perhaps another illustration will make this point more clearly. The thirteenth-century logician William of Sherwood cites two arguments as incorrect; he presents them as instances of Ignorance Regarding Refutation (*Ignoratio Elenchi*):

(1) Socrates is naturally pious, but he is not absolutely pious; therefore he is both pious and not pious.

(2) Socrates is running at time *a* [*currit in a*] and he is not running at time *b*; therefore, he is both running and not running.

Notice that in both these arguments there is subject-matter overlap, or relatedness, between premisses and conclusion, yet both arguments are clearly incorrect. Indeed, both are sophismatical, or fallacious, arguments from premisses that are possibly true to a conclusion that can't be true.

So subject-matter overlap is not enough to rule out fallaciousness. To be a correct argument meeting reasonable standards that will avoid *ignoratio elenchi*, there must be subject-matter overlap; also, the argument must not move from true premisses to a false conclusion. Thus there are two requirements: the relatedness of premisses and conclusion; and the truth of the conclusion on the condition that the premisses are true. Hence a *relatedness conditional, p → q*, must incorporate both requirements: that (1) p is related to q, and that (2) it is not the case that p is true and q is false.

In short, Senator Martin's argument fails not through failure of subject-matter overlap; it fails because the premisses fail to imply the conclusion, either deductively or inductively. It is neither impossible nor improbable that a grimy city might have hard-working and tolerant citizens.

Even so, one might persist with the suggestion that an argument might fail by reason of lack of subject-matter overlap. Let x be any city.

Premiss 1:   For all x, x has fine schools.
Premiss 2:   For all x, x has hard-working, tolerant citizens.
Conclusion: For all x, x is not a grimy city.

To be sure, this argument does fail to have significant subject-matter overlap; also, it is a bad argument.

So perhaps there is a second explanation of what is fallacious about Senator Martin's argument. It is a true red herring after all; it isn't merely a failure of implication not specifically due to the lack of a common subject-matter.

As we have said, the traditional Aristotelian fallacy of ignorance of refutation was not restricted merely to instances of failure of subject-matter overlap. Indeed, the Aristotelian tradition of the "topics", which was so influential in mediaeval logic, distinguished numerous different kinds of connections, or "topics", which could relate the premisses and conclusion of a correct argument. Boethius followed Cicero's conception of a topic as *sedes argumenti* (a seat of argument), and described it as "that from which a fitting argument may be drawn for a proposed question".[10] For these earlier thinkers, a topic functions primarily as a way of *finding new arguments*; but for later writers, such as Abelard and Ockham, the topic became the *inferential basis of a conditional proposition*.[11]

Abelard claims that the topics show the inferential force (*vis inferentiae*) of all conditionals. According to one commentator, Abelard uses the topic as an "inference-warrant" that can accommodate formal as well as non-formal inferences.[12] Abelard gives two examples. "If it is man, it is animal." According to Abelard, this is a correct (good and necessary) inference. But "If it is stone, it is animal" is not a good argument, because it lacks what he calls the relation of the Topical Difference of Species.

However, it is clear that the relation of genus to species was not the only topical relation recognized by the mediaevals. In fact, the diversity of the dialectical topics subdivided the kinds of conditionals they recognized to be correct into many classifications. For example, Peter of Spain lists some twenty-one topics, including part-whole relations, attributions of place and time, causation, similarity, authority, adverbial modification, and so forth.[13] What is common to everything on his list is that a topic is always a relation that warrants an inference.

This tradition of the topics suggests that it is no simple matter to apply formal theories of the conditional to the wide variety of incorrect reasoning involved in many informal fallacies and practical arguments. Theory of topics appears to involve many different kinds of relations that link the antecedent to the consequent in conditionals. Subject-matter overlap is *one* such relation that helps to clarify *one* sense of "failure of relevance" central to an understanding of the modern conceptions of the *ad populum, ad hominem*, and similar fallacies.

Let us now try to formulate a *general theory of relatedness* to accommodate our requirements and still allow for a good deal of flexibility in application to specific arguments.

Many of those who are drawn to informal logic see it as a subject that shows pedagogical promise in teaching students how to handle the practicalities of evaluation of argumentation that have to do with failures of topical relevance, and other factors related to meaning. Yet the clear guidelines offered by a formal theory are not to be excluded, either.

Topic overlap and *deductive correctness of argument* are often bound together when we evaluate arguments. It seems, therefore, most unfortunate for dialectic that the classical deductive notion of entailment does not take the factor of topic relevance into account at all. But could a concept of entailment be constructed that is sensitive to topic overlap, as well as to truth-values? The answer is yes.

Deductive logic can be extended in this direction. We shall devote the next chapter to showing how it can be done, by developing deductive logic more fully as a formal system. Then, we will see how the seemingly "informal" aspect of topic overlap can be taken into account within the context of formal logic.

Notes

1 Good *deductions*, but not perfectly blemish-free arguments. Remember, they are circular.

2   There is *exactly one* exception to this rule. Can you identify it?

3   Here the only constituent elements whose meaning need not be taken strictly into account are the two occurrences of the phrase, "the shirt", for *anything* that is red is *also* coloured.

4   Augustus De Morgan, *Formal Logic* (London: Taylor and Walton, 1847), p. 246.

5   *Natural* theology is that branch of theology that seeks to discuss the properties of divinity by inference from various facts of the natural world.

6   Richard B. Angell, *Reasoning and Logic* (New York: Appleton-Century-Crofts, 1964), page 442.

7   Richard Whately, *Elements of Logic* (New York: William Jackson, 1836) p. 195.

8   The examples are from Josef L. Altholz, "Truth and Equivocation: Liquoi's Moral Theology and Newman's Apologia," *Church History* LXIV (1975): 73–84. And there are two interesting articles that deal with how to define "lying": (1) Frederick A. Siegler, "Lying," *American Philosophical Quarterly* 3 (1966): 128–136; (2) Roderick Chisholm and Thomas D. Feehan, "The Intent to Deceive," *Journal of Philosophy* 74 (1977): 143–159.

9   A very early version of this paradox comes from the epistle of St. Paul (Titus 1: 12–13); the version we present here is from Alfred Tarski, *Logic, Semantics and Metamathematics* (New York: Oxford University Press, 1956).

10   Boethius, *De Topicis Differentiis*, trans. Eleonore Stump (Ithaca, N.Y.: Cornell University Press, 1978). He was a mediaeval philosopher (470–525 AD) who was strongly influenced by Aristotle and Cicero.

11   *Peter Abelard* is the French philosopher who flourished in the first third of the twelfth century, and is well-known to romantic historians for a passionate and tragic love-affair with the gifted Héloïse. *William of Ockham* is the very considerable English philosopher who flourished in the early fourteenth century. Ockham was an influential logician, and his theological writings rivalled and challenged those of Thomas Aquinas.

12   Otto Bird, "The Formalizing of Topics in Medieval Logic," *Notre Dame Journal of Formal Logic* 1 (1960): 138–149.

13   *Peter of Spain* flourished in the mid-thirteenth century. He is the author of an influential textbook on logic; as John XXI, he was the first and only Portuguese Pope. Less than a year after his papal election, the ceiling of his library at Viterbo fell and killed him.

Chapter Nine
Formal Systems

In this chapter we want to show how logic can be organized into formal systems; to formulate the unequivocal conventions of correctness; and to fix meanings precisely. To do this, we develop the rudiments of two simple formal systems. The first is our old friend, classical truth-functional logic. The second system is modelled on the first, but fixes the meanings of some of the constants in a different way, in order to explore the idea of *connections of subject-matters between pairs of propositions in an argument.*

1. An Elementary Formal System

By an elementary formal system, we mean a system of logical forms that represent the simplest parts of deductive logic. As we know now, a logical form is a configuration of sentence variables and sentence constants in which uniform replacement preserves deductive correctness. Our first system, a classical elementary system, must therefore accomplish three things:

(1) It must provide a means of constructing logical forms. *This is the construction requirement.*

(2) It must clearly define or specify the logically interesting properties that we would want these forms to be tested for. (We have already seen that deductive correctness is *not* the only logically interesting property to be investigated.) *This is the definition requirement.*

(3) And it must provide flawless mechanisms for detecting the presence or absence of these various logically interesting properties in the logical forms examined. *This is the decision requirement.*

In developing our logical system there are three requirements we need to keep in mind:

> The construction requirement
> The definition requirement
> The decision requirement

Let us call our logical system "S". We turn now to our three requirements.

(1) The construction requirement for the system S.

First, we need a stock of sentence variables:

$$A, B, C, D, A_1, A_2, \ldots, A_n, \ldots$$

It is worth noticing that we have given ourselves rather a lot of sentence variables – in fact, an infinity of them. By so doing, we absolutely guarantee that there is no argument, however complex, that we cannot analyze just because we would run out of variables in constructing its logical form.

We also make use, in our system S, of five *sentence constants:*

> ⌐, pronounced as "not-"
> V, pronounced as "or"
> ∧, pronounced as "and"
> ⊃, pronounced as "if . . . then"
> ≡, pronounced as "if and only if"

We appreciate that in this formal system the meaning of some of the constants is not the same as the corresponding words or phrases in English. The most pronounced deviation, as we saw in Chapter Three, is the hook, ⊃. For our first system of logic, the hook remains a deviation; in our second formal system we will see how to correct this deviation.

Another item we will use to construct our sentences is the parentheses, which are used as punctuators. We are already familiar with the use of parentheses from Chapter Three.

The construction requirement for the system S also necessitates that we make use of two sets of *construction rules*. The one set of construction rules tells us how to construct *formulas*, and the other set of construction rules tells us how to construct *deductive logical forms*. Let us begin with the construction rules for formulas.

CONSTRUCTION RULES FOR FORMULAS OF S:
1. Any sentence variable standing alone is a formula of the system S.
2. Suppose that you have some formulas, any two formulas you like. Give them the names "A" and "B". Then the following are also formulas of the system S:
   (⌐A)
   (A V B)
   (A ∧ B)
   (A ⊃ B)
   (A ≡ B)
3. Nothing else is a formula of S.

We add the convention that the outmost parentheses in a formula may be dropped. For example, in the formula (A V B) ∧ C, the parentheses are needed in order to distinguish between the two sets of elements: A, B are in one set, C another. But in the simple formula (A ∨ B), the outer parentheses may be dropped, at our convenience.

Now let us examine the rules for deductive logical forms in S.

CONSTRUCTION RULES FOR DEDUCTIVE LOGICAL FORMS OF S:

1. Take any finite number of formulas A, B, C, and so on. Select exactly one of them – say, D – and flag it with the *conclusion indicator* "∴". Thus:

    ∴ D

    Call D a "conclusion".

2. Above your conclusion, write *at least one* formula for each new line. Repetitions are allowed. Call these "premisses". The result of this operation is a deductive logical form of S.

3. Nothing else is a deductive logical form of S.

Example

By means of the construction rules for formulas, we can determine whether a thing is, or is not, a formula of S. How would we decide whether an expression like the one below is a formula of S?

$$¬(A ∨ B) ⊃ ((¬A ∧ ¬B) ≡ A)$$

We can decide by applying the construction rules of S, starting from the innermost formulas and working outwards. First, we know that all occurrences of A and B are formulas. Second, we know, by the construction rule for ¬, that the symbol ¬ in front of a formula produces a formula. So we know that ¬ A and ¬ B are formulas:

$$¬(A ∨ B) ⊃ ((\underline{¬A} ∧ \underline{¬B}) ≡ A)$$

We also know that putting a ∨ or a ∧ between two formulas yields a formula, so we know that we have several other formulas:

$$¬(\underline{A ∨ B}) ⊃ ((\underline{¬A ∧ ¬B}) ≡ A)$$

Proceeding to apply the rules, we can eventually see that the whole expression is a formula:

$$(\underline{A ∨ B}) ⊃ ((\underline{¬A ∧ ¬B}) ≡ \underline{A})$$

Examples

The following four sequences are all deductive logical forms of S.

(1) *A*
    ———
    ∴ A

(2) *A ∧ B*
    ———
    ∴ B

(3)   A
      A
      B ⊃ C
      D
      ――――――
      ∴ E ∨ F

(4)   A ⊃ (B ∨ ⅂ C)
      ⅂⅂ D ∨ (B ∧ A)
      ――――――――――
         ∴ C

Exercise

(a) Determine whether each of the following expressions is a formula of
     S.

     1. (A ⊃ B) ∨ (C ⊃ D ∨ A)
     2. (A ⊃ B) ∨ (B ⊃ C) ⊃ (A ⊃ C)
     3. ((A ⊃ ⅂ B) ∧ (B ⊃ ⅂ C)) ∨ D
     4. (A ∧ B ∨ C)
     5. ⅂ ((⅂ A ⊃ ⅂ B) ⊃ (⅂ B ⊃ ⅂ A))

(b) Determine whether each of the following expressions is a deductive
     logical form of S.

     1. A ⊃ B
         A ∨ B ∧ C
         ――――――――
            ∴ B

     2. B
         B ∧ C
         ―――――
         ∴ A ∧ B

     3. A
         B ∨ C
         ―――――
         ∴ B
         ∴ C

(2)   The definition requirement for the system S.

Now that we have the means of constructing the formulas and deductive
logical forms of our system S, we must specify their logically interesting
properties.

When joined in appropriate ways with one or more parenthesized
formulas, each constant produces a *new* formula. As we can appreciate,
formulas are like sentences, except that their meanings are suppressed.
Since formulas are used to make up deductive logical forms, it is
convenient to stipulate that they are either *true* or *false*, but not both. As
we said in Chapter Two, a logician calls the truth or falsity of a sentence
its *truth-value*. We shall retain that practice here.

Let us pause and renew our bearings.

1. The constants of the system S are united, by parenthesized formulas, to form new formulas.
2. Every formula has a truth-value.
3. The formula to which a constant applies has a truth-value; the new formula, which results, also has a truth-value.
4. In our system, constants are the elements that have meaning. Thus the truth-value of a resulting formula depends on two things:
   (a) the truth-values of the constituent formulas
   (b) the meanings of the constants.

Our constants are defined by truth-tables, just as in Chapter Three, so that the truth-value of each formula is uniquely determined by the truth-values of the constituent formulas, and by nothing else. Every formula of S has its own unique truth-table, and that truth-table is fixed by the definitions of the constants, ⌐, ∧, ∨, ⊃, and ≡.

Exercise

As a reminder of how truth-tables work, construct truth-tables for the following formulas of S:
(a) $(A ⊃ B) ∨ (⌐B ⊃ ⌐A)$
(b) $((A ∧ B) ⊃ C) ⊃ (A ⊃ C)$

A formula's truth-table tells you, in an absolutely *reliable* way, *mechanically* and in a *finite* time, the truth-values a formula will have, whatever the truth-values of its constituent formulas. We can conclude that requirement 3, the decision requirement for S, is met.

2. Some Logically Interesting Properties of System S

Certain formulas will be true no matter what truth-values we give to the variables in them. Recall the truth-table for A ∨ ⌐ A:

A	⌐ A	A ∨ ⌐ A
T	F	T
F	T	T

A formula that is *always* true is *necessarily* true. Let us call such a formula a *truth-functional tautology*. Many formulas of S are sometimes true and sometimes false – for example, A ∨ B. Let us call these *contingent formulas*. Others are always false – for example, A ∧ ⌐ A.

A	⌐ A	A ∧ ⌐ A
T	F	F
F	T	F

Let us call those formulas that are always false *inconsistent formulas* or *contradictions* in S.

It is easy enough to see that the properties of truth-functional redundancy, contingency, and contradictoriness are of interest to logicians. Logic is an instrument to help us get at the truth of things. Thus if we are able to determine that a given formula is a truth-functional tautology, we will have guaranteed for ourselves knowledge of at least a small part of the truth of things. On the other hand, if we can discover that a given formula is a truth-functional contradiction, we would know that that formula does not serve the cause of truth. In the case of a formula whose truth-table shows it to be contingent, we discover that the formula is true under certain circumstances; moreover, we discover what those certain circumstances are, namely, when its constituent variables themselves have truth-values that, according to the truth-table, make the whole formula come out true.

One small point: in working with system S, it is worthwhile to note that some parentheses may be eliminated in order to simplify our notation when we are dealing with a number of conjunctions nested together. If you construct a truth-table for A ∧ (B ∧ C) and one for (A ∧ B) ∧ C, you will see that the last columns of the two tables are identical. If you have conjunctive formulas nested together in this way, it doesn't matter where you put the parenthesis. You could simply write A ∧ B ∧ C, and there would be no ambiguity. In other words, if you say "A and B and C", it doesn't matter whether you punctuate the sentence or not. There is no difference between "A and B, and C" and "A, and B and C".

But in looking at deductive logic, we are interested primarily in questions of the deductive correctness and incorrectness of *arguments*. Our emphasis, therefore, should fall upon argument forms rather than upon formulas. Are there any logically interesting properties of argument forms, which our truth-tables help us to specify? The answer is decidedly yes.

To illustrate the point, consider any argument form you like. For example:

$$A, \therefore B$$

Recall that A stands for all the premisses of the argument and B for the argument's conclusion. We will now form what logicians call the *conditional formula*. Each argument has a corresponding formula, which is located very simply, in two steps.

Step 1:   Collect all the formulas which make up the group of premisses, A, and form their conjunction.

Example. Let us say that A contains the three formulas C, C ⊃ D, and E. We would form their conjunction as follows:

$$C \wedge (C \supset D) \wedge E.$$

185

Note that we are dropping unnecessary parentheses around the conjunctions.

Step 2: Take this conjunction and put parentheses around it. Take the conclusion and now form the conditional formula with the conjunction as antecedent and the conclusion as consequent. *Example.* Let our conclusion be the formula F. The argument with which we began is the deduction:

$$C$$
$$C \supset D$$
$$E$$
_____
$$F$$

The conditional formula corresponding to this argument is (C ∧ (C ⊃ D) ∧ E) ⊃ F.

We can now assert a proposition of fundamental importance to the deductive logic of sentences:

> An argument form, in S, is deductively correct if and only if its corresponding conditional is a truth-functional tautology.[1]

Thus deductive sentence logic has what may be called a "truth-table test" for deductive correctness. Every logical form that can be expressed in the system S is subject to this test. As we have already mentioned, the test is perfectly reliable. It may be used in a purely mechanical way, and takes only finite time. Procedures for detecting the presence and absence of logically interesting properties are called *effective procedures* for such properties. Properties for which effective procedures exist are called *decidable properties*.

In the system S, there are five decidable properties:

1. The deductive correctness of a logical form
2. The deductive incorrectness of a logical form
3. The truth functional redundancy of a formula
4. The contingency of a formula
5. The truth functional contradictoriness of a formula

Exercise

Construct the conditional formula corresponding to the following arguments:

(a) A ∨ B
    A ⊃ C
    B ⊃ D
   _____
   ∴ C ∨ D

(b) ⏋ B
    ⏋ C
    B ⊃ C
    D ∨ E
   _____
   C ⊃ ⏋ E

A brief reprise. In the chapters beginning this book, we touched on a number of logically important notions; it is now time to tidy up our understanding of them. In Chapter Three, we spoke of correct or warranted *inference*. In the same chapter, we spoke of *entailment*. In several places, we have spoken of *deductive correctness*. And in this chapter, we have met with the notion of *truth functionally tautologous conditional formulas*. Do these various notions interrelate with one another?

They do indeed interrelate. According to formal system S, the interconnections are three: (1) An inference of a conclusion from premises is warranted, in system S, if and only if the argument composed of those premises and that conclusion is deductively correct. (2) An argument, p, $\therefore q$, is deductively correct in S if and only if the conjunction of its premises entails its conclusion. (3) A formula, A, entails a formula, B, in S if and only if the S formula A ⊃ B is a truth functional tautology.

The interconnections claimed by the system S are incredibly important. For all that is necessary for the determination of the correctness of deductions, of logical entailments, of warranted inferences, and of the truth functional redundancy of conditionals is the effective device of the truth-table test for the formulas of S. Accordingly, the notions of warranted inference, entailment, and deductive correctness are all decidable in S. What is more, the logical structure of any of these three concepts will be basically the same. For example, the logic of entailment could be expected to be essentially the same as the logic of warranted inference.

Some Examples of Formulas and Logical Forms Displaying the Logically Interesting Property of System S

For the purposes of this chapter, it does not matter in the slightest whether we are right in thinking that system S overstates the interconnections among the four concepts that we have just been considering. Let us turn to some prominent examples that illustrate the most important of the logically interesting properties of system S. Our examples will also reflect the pattern of interconnections claimed for them by system S.

If this formula is tautologous,	this deduction is correct,	and this inference is warranted.
((A ⊃ B) ∧ A) ⊃ B	A ⊃ B A ——— ∴ B	From A ⊃ B and A, it is warranted to infer B.

$((A \supset B) \wedge \urcorner B) \supset \urcorner A$

$A \supset B$
$\urcorner B$
_____
$\therefore \urcorner A$

From $A \supset B$ and $\urcorner B$, it is warranted to infer $\urcorner A$.

$A \supset (A \vee B)$

A
_____
$\therefore A \vee B$

From A, it is warranted to infer $A \vee B$.

$((A \vee B) \wedge \urcorner A) \supset B$

$A \vee B$
$\urcorner A$
_____
$\therefore B$

From $A \vee B$ and $\urcorner A$, it is warranted to infer B.

$(A \wedge B) \supset A$

$A \wedge B$
_____
$\therefore A$

From $A \wedge B$, it is warranted to infer A.

$((A \supset B) \wedge (B \supset C)) \supset$
$(A \supset C)$

$A \supset B$
$B \supset C$
_____
$\therefore A \supset C$

From $A \supset B$ and $B \supset C$, it is warranted to infer $A \supset C$.

Exercise

By means of six truth-tables, show that each of the six conditionalized formulas in our example is a tautology.

Logically Equivalent Forms of S

We have already met with the notion of entailment. A sentence or formula, A, entails a sentence or formula, B, when the truth of A guarantees the truth of B. Sometimes, however, the truth of A guarantees the truth of B *and* the truth of B guarantees the truth of A. In other words, sometimes A entails B and B entails A; that is, entailment is *reciprocal*. Propositions that reciprocally entail one another *are guaranteed always to have the same truth-value*. And propositions that always and unavoidably have the same truth-value are said to be *logically equivalent*.

The relation of logical equivalence can, in fact, be defined within the system S. If you think about it a bit, any two formulas of S – say the formulas A and B – will be logically equivalent to one another when the biconditional sentence A ≡ B is a truth-functional tautology in S.

The idea of logical equivalence is a powerful and useful one. In particular, the system S always permits that in any deduction in S, any formula may be replaced by a logically equivalent formula. Fortunately, we need never be in doubt about whether formulas are logically equivalent: we need only construct the appropriate biconditional formula and then apply the truth-table test of redundancy. If the biconditional sentence tests out to be a tautology, then the two constituent sentences are logically equivalent. If the biconditional formula does not test out to be a tautology, then the constituent formulas are not logically equivalent. Though the truth-table test of redundancy always enables us to discover the presence or absence of logical equivalence in S, nevertheless it is useful to have some general idea of formulas that are logically equivalent. Accordingly, we will list some of the more important kinds of equivalent formulas.

Some Logically Equivalent Formulas of S.

(1a)   ˥ (A ∨ B) and ˥ A ∧ ˥ B
(1b)   ˥ (A ∧ B) and ˥ A ∨ ˥ B
These formulas are declared equivalent by *De Morgan's Laws of S*.

(2a)   A ∨ B and B ∨ A
(2b)   A ∧ B and B ∧ A
These formulas are declared equivalent by the *Commutative Laws of S*.

(3a)   A ∨ (B ∨ C) and (A ∨ B) ∨ C
(3b)   A ∧ (B ∧ C) and (A ∧ B) ∧ C
These formulas are declared equivalent by the *Association Laws of S*.

(4a)   A ∧ (B ∨ C) and (A ∧ B) ∨ (A ∧ C)
(4b)   A ∨ (B ∧ C) and (A ∨ B) ∧ (A ∨ C)
These formulas are declared equivalent by the *Distribution Laws of S*.

(5)   A and ˥ ˥ A
These formulas are declared equivalent by the *Double Negation Laws of S*.

(6)   A ⊃ B and ˥ B ⊃ ˥ A
These formulas are declared equivalent by the *Contraposition Law of S*.

(7)   A ⊃ B and ˥ A ∨ B
These formulas are declared equivalent by the *Conditional Law of S*.

(8)   A ≡ B and (A ⊃ B) ∧ (B ⊃ A)
These formulas are declared equivalent by the *Biconditional Law of S*.

(9)   (A ∧ B) ⊃ C and A ⊃ (B ⊃ C)
These formulas are declared equivalent by the *Exportation Law of S*.

(10)   A and A ∨ A
         A and A ∧ A
These formulas are declared equivalent by the *Repetition Laws of S*.

Exercise

By means of truth-tables, show that the pairs of formulas in (1b), (6), (7) and (9) are logically equivalent pairs.

3. The System S as a Logical Theory

In S, connectives do not always have the same meaning as their English counterparts. Yet obviously we do our reasoning and conduct our arguments in English. So why should we permit the theory of deductive logic to be formulated in S? This is a good question, and a fair one.

Logicians have an answer. Logic is a *theory*.

(1) A theory is a systematic and orderly organization of what is already known of a given subject-matter, whether it be physics, psychology, household economics or logic. What is already known may have been known only in a common-sense sort of way and without much recognition of underlying principles. A theory, however, not only preserves what is already known: it also articulates and exposes underlying principles.

(2) A theory also discovers new knowledge. In particular, by the orderly and systematic exposure of underlying principles, a theory discovers new implications of those principles and new applications of them, which may not have been apparent at the common-sense level.

(3) In order to achieve the necessary orderliness, a theory usually involves reference to somewhat abstract and idealized concepts. For example, ordinary theoretical physics talks about point masses and frictionless surfaces. A point mass is a quantity of matter just large enough to occupy a geometrical point, and a frictionless surface is one on which, except for the application of external forces, a body's velocity would never alter. Strictly speaking, there are no point masses and there are no frictionless surfaces. These are idealized abstractions.

Nevertheless the physics of point masses and frictionless surfaces is very systematic and extremely highly organized. And because these idealized abstractions are quite similar to actual particles and to slippery surfaces in nature, the laws of theoretical physics have a tendency to apply to the real world. Perhaps they do not apply exactly, but certainly they apply in such a way as to give us a very good idea of how nature operates.

In the *theoretical* branches of physics, the term "particle" ("point mass") does not mean exactly the same as the term "particle" in ordinary, common-sense English, but the two words have a quite similar meaning. And in theoretical physics, "frictionless surface" does not mean just the same as "frictionless surface" in ordinary, common-sense English. In general, a theory is permitted to change or refine the meaning of common-sense terms and expressions, provided that

(i)   the refined meanings are quite similar to the common-sense meanings, and

(ii) the refined meanings make it easy to systematize and organize the subject-matter of the theory. These refined meanings are sometimes called explications of their ordinary, common-sense counterparts. In a theory, then, explications (which involve refinements in the meaning of common-sense terms) may be said to denote the abstract and idealized concepts so essential to the powerful orderliness of the theory in question.

Let us take a useful example from our theory S. We have already seen that the hook, $\supset$, does not mean the same thing as "if . . . then", its ordinary, common-sense English translation. You cannot always tell whether an "if . . . then" sentence is true, *even if you know the truth-value of its antecedent and consequent.* In S, on the other hand, we always know the truth-value of a $\supset$ formula once we know the truth-values of the antecedent and the consequent. So there is a *difference* between "if . . . then" and $\supset$. Does the refined meaning of $\supset$ make it easier to systematize and organize the subject-matter of deductive logic? It certainly does. In the refined English of S, it is *always* possible to determine whether a hook formula is a tautology; therefore, it is always possible to tell whether the deduction to which it corresponds is correct.

Even so, this theoretical refinement is allowable only if $\supset$ and "if . . . then" have a reasonably similar meaning. To some extent, they do: both $\supset$ and "if . . . then" obey the fundamental law of implication.

LAW
A true proposition never implies a false proposition.

In the truth-table for $\supset$, if formula A is true and formula B is false, $A \supset B$ is always false. And in common-sense English the same also holds true.

Thus we can see that the use of the hook, $\supset$, in place of the common-sense "if . . . then" is, to a certain extent, justified. It allows for greater organization, and it does not stray too far from the meaning of "if . . . then".

In summary, then, we can say that a theory

1) preserves what is already known;
2) articulates underlying principles;
3) organizes underlying principles in such a way that new knowledge can be discovered;
4) may use a refined vocabulary of special terms, which are explications of certain ordinary, common-sense terms; such applications denote abstract and idealized concepts.

It is a simple matter to ascertain that S is indeed a theory.

However, we have known all along that the system S achieves its theoretical simplicity at the expense of what it ignores – namely, any question of the connections between A and B apart from the individual truth-values of A and B. The resulting vulnerability to the *ignoratio elenchi* appeared, for example, in the acceptance in S of the formula A ⊃ (A ∨ B) as a tautology. The fact that this formula is a tautology means, of course, that from A we are warranted to infer A ∨ B *for any* A *and* B *you like,* even if they are totally unrelated. From the statement "Bananas are yellow," we can deductively infer "Either bananas are yellow or it will rain on Tuesday." How could we extend or modify S to cope with this limitation? We do it by constructing a new system, NS.

4. The System NS: Its Elements

System NS starts out the same way S did – it has sentence variables and parentheses. But the constants of NS are defined somewhat differently. The construction rules of NS are the same as those in S, except that instead of ⊃, we will have a new conditional, → (arrow), and some of the other constants will be changed somewhat as well. Additionally, we will have in NS a relation between propositions *r* (*p, q*) that holds if *p* is related to *q*, but fails if *p* and *q* are not related.

To answer the problem of the *ignoratio elenchi*, we can explain *r* (*p, q*) as follows. Select a set of topics in a given argument, and call this set T. T can be any set of topics you clearly agree upon. Any proposition (*p*) in the argument may contain a number of different *subject-matters* that are elements of T. For example, suppose T contains "bananas", "yellow", and a number of other topics. The proposition "Bananas are yellow" will contain two subject-matters from T, namely "bananas" and "yellow". The subject-matter, then, is what the proposition is *about*.

We will say that *p* is related to *q* if there is at least one common element of subject-matter between *p* and *q*. For example: if *p* is "Bananas are yellow" and *q* is "Malaria makes your skin yellow", then obviously *p* and *q* are related: both contain the common subject-matter "yellow".

We can perhaps better understand what is meant by *r* if we see that it has three characteristic properties.

Reflexivity: r (p, p). That is, *p* is always related to itself. This is reasonable, because any proposition will clearly have some overlapping subject-matter with itself.

Symmetry: r (p, q) if, and only if, *r (q, p).* This is reasonable, because if *p* shares some topics with *q*, then *q* will also share some topics with *p*. Sharing subject-matters works both ways.

Failure of transitivity: if *r (p, q)* and *r (q, s),* it need not follow that *r (p, s).* Here is an example. Suppose *p* overlaps in subject-matters with *q*, and suppose that *q* overlaps in subject-matters with *r*:

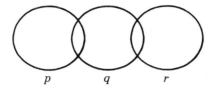

Still, it need not follow that p shares any overlapping subject-matters with r, as the diagram indicates. They might not have any common topics at all. For example, "Bananas are yellow" shares a topic with "Woods hates bananas", and "Woods hates bananas" shares a topic with "Woods has two daughters". But "Bananas are yellow" does not seem to be related to "Woods has two daughters".

These three characteristics give us some idea of how to understand r. There might be other ways to define r, but these characteristics suit the approach of subject-matter relatedness.

Exercise

Do the following relations have the properties of reflexivity, symmetry, and transitivity? Or do some of these properties fail? Which relations have which properties?
(a) x is north of y (for locations x and y)
(b) x loves y (for persons x and y)
(c) x kills y (for persons x and y)
(d) p entails q (for propositions p and q)
(e) the subject-matter of p includes the subject-matter of q (for propositions p and q)

Now we have some idea of what r means. We can see how we need to define the conditional, $\rightarrow$, in NS. If p is related to q, then the truth-table definition for $\rightarrow$ can be the same as the truth-table definition for $\supset$ in S. But if p and q are unrelated, then we will say that $p \rightarrow q$ is false. We define $\rightarrow$ this way because we want to require that "If p then q" fails to obtain if p and q are unrelated. If p and q are clearly unrelated, and an *ignoratio elenchi* has occurred, then the conditional must fail to hold. So, in order for "If p then q" to be true, p *must* be related to q. Accordingly, the truth-table for $\rightarrow$ is defined:

p	q	$r\,(p, q)$	$p \rightarrow q$
T	T	T	T
T	F	T	F
F	T	T	T
F	F	T	T
T	T	F	F
T	F	F	F
F	T	F	F
F	F	F	F

The idea is simple. If p and q are related, the truth-table for $\rightarrow$ is the same as the truth-table for $\supset$. If p and q are not related, $p \rightarrow q$ always comes out false.

The other constants are also straightforward. In NS, ⌐ is defined as it is in S. Reason: negation does not seem to be influenced by subject-matters. In NS, as in S, ⌐p always has the opposite truth-value of p, regardless of what p is about. Conjunction and disjunction can be defined with or without relatedness, as you prefer. But it seems natural to think that "and" does not usually require relatedness, while "or" most often does. "Either bananas are yellow or $2 + 2 = 4$" does not sound right. But we often conjoin $p \wedge q \wedge r$ and so on in a list, where these propositions may not share subject-matters at all.

For our purposes, therefore, we will define $\wedge$ just as in S. But we define $p \vee q$ as follows: $p \vee q$ is true if p is true or if q is true, and if p is related to q. Thus $p \vee q$ is false if both p and q are false, or if they are unrelated. The truth-table for $p \vee q$ will look like the truth-table for $p \rightarrow q$, except that the final column for the first four rows will read: TTTF. If you object to this way of proceeding, it will be possible to define $\vee$ and $\wedge$ either way you like. Finally, $p \longleftrightarrow q$ (the relatedness bi-conditional) is defined as $p \rightarrow q$ and $q \rightarrow p$. So $\longleftrightarrow$ will require relatedness too.

In order to satisfy the construction requirement for NS, we have to see how the bigger formulas containing the constants ⌐, $\rightarrow$, $\vee$, $\wedge$, and $\equiv$ are made up from the single propositions, p, q, r, and so on. To do this, we have to decide how the complex formulas (made up of the constants and variables) are related to each other. Given a formula A and a formula B, no matter how complex they might be, we need to decide how they are related. For example: if A is the proposition p, and if B is the proposition $q \rightarrow s$, we don't have any way of automatically determining whether A is related to B. It could be that p is related to q, but that doesn't tell us whether p is related to $q \rightarrow s$. We might want to say that p is not related to $q \rightarrow s$ unless p is related to both q and s. Or we might allow that p is related to $q \rightarrow s$ if p is related to q or s. How to decide?

We will say that any two formulas of the form A and B $\rightarrow$ C are related if A is related to B *or* if A is related to C. This seems reasonable. Say we have a conditional such as "If bananas are yellow then Einstein discovered relativity." As a relatedness conditional, it is false. But should we say that "Einstein was absent-minded" is related to it? Yes, it would seem, because the latter proposition is related to one constituent of the conditional.

The construction rule for $\vee$ is the same as that for $\rightarrow$. A is related to B $\vee$ C if, and only if, A is related to B or A is related to C.

Exercise

1. If A is related to B, and B is related to C, but A is not related to

C, determine which pairs of the following formulas are related to each other in NS.

  (a) A ∨ B
  (b) B → C
  (c) A → (B → C)
  (d) ⊐ A ∨ B
  (e) (A ∧ B) → C
  (f) A → A

2. Suppose A is related to B; B is not related to C; A is not related to C; suppose finally that A and B are true but that C is false. Determine the truth-values of each of the following formulas of NS.

  (a) A → B
  (b) B → C
  (c) A → (B → C)
  (d) A ∨ B
  (e) A ∧ ⊐ B
  (f) A ∧ (B ∨ C)
  (g) B → (B → C)
  (h) (A → B) ∨ (B → C)

As in S, it is true in NS that the truth-value of any formula depends on two things: (i) the truth-value of the constituent formulas, and (ii) the meaning of the constants. The only difference between S and NS is that in NS, two of the constants, → and ∨, are defined in a different way. But NS also meets the construction requirement, because we can always determine, by procedures exactly similar to the construction rules for S, whether something is a formula of NS. And a similar property to truth-functionality in S holds in NS: given the truth-values of the constituent propositions in any formula of S, and given the relatedness relations that obtain between these constituent propositions, we can always determine the truth-value of that formula. In NS, just as in S, the truth-table is a method that can always tell you, reliably and mechanically and in a finite time, the truth-values a formula will have, given the truth-values and relations of its constituent formulas. So NS meets the decision requirement.

5. Some Logically Interesting Properties of NS

Certain formulas will be true in NS no matter what truth-values we give to them, and no matter what relatedness relations the constituent propositions stand to each other in. Consider the truth-table for A ∨ ⊐ A:

A	⊐ A	A R ⊐ A	A ∨ ⊐ A
T	F	T	T
F	T	T	T

We adopt the convention of writing "A R B" for "A is related to B" where the formulas are related. We had used "*r*" to represent relatedness of the constituent propositions *p, q, r,* and so on before we considered the case of formulas generally, where formulas A, B, C, . . . may contain complex expressions like A → B and so forth.

Now examine the truth-table above. We know already that A R ⌐ A is always true, because A is always related to itself (remember the property of reflexivity from the last section). And we know that a formula, A, is related to the negation of a formula, say ⌐ B, if and only if A is related to B. Thus A is always related to ⌐ A, regardless of the truth-value of A. Therefore A ∨ ⌐ A will always be true. As long as A is related to ⌐ A, then A ∨ ⌐ A has the same values in NS as it had in S. We shall call such a universally true formula in NS a *relatedness tautology*.

If a formula of NS is sometimes true and sometimes false, we will call it a *relatedness contingent formula*. Formulas that are always false we call the *relatedness contradictions* or *relatedness inconsistent formulas*. For example, A ∧ ⌐ A is inconsistent in NS. Because ∧ does not require relatedness at all in NS, the truth-table for A ∧ ⌐ A is the same in NS as it was in S (see page 184).

Do systems S and NS differ in classifying formulas as tautologies? Yes. Some formulas that are tautologies in S are *not* tautologies when expressed as formulas of NS. Consider the S truth-table for A ⊃ (B ⊃ A).

A	B	B ⊃ A	A ⊃ (B ⊃ A)
T	T	T	T
T	F	T	T
F	T	F	T
F	F	T	T

Every row of the last column is a true; thus the formula is a tautology in S. But consider the relatedness truth-table for the parallel formula of NS, A → (B → A):

	A	B	A R B	B → A	A → (B → A)
(1)	T	T	T	T	T
(2)	T	F	T	T	T
(3)	F	T	T	F	T
(4)	F	F	T	T	T
(5)	T	T	F	F	F
(6)	T	F	F	F	F
(7)	F	T	F	F	T
(8)	F	F	F	F	T

There are two cases where A → (B → A) is not true: rows (5) and (6). In row (5) we have it that A is not related to B. Therefore B → A must be false.

Now look at A → (B → A). A is related to (B → A) because A is related to A (remember that a formula, A, is related to a conditional B → C if A is related to B *or* if A is related to C). Therefore A → (B → A) is false only if A is true and B → A is false. But in rows (5) and (6), A is indeed true and B → A is false. Thus the whole formula A → (B → A) is false in rows (5) and (6). So clearly A → (B → A) cannot be a relatedness tautology. As the truth-table shows, it is a contingent formula in NS.

Exercise

1. Determine whether ˥ A → (A → B) is a relatedness tautology, a relatedness contingent formula, or a relatedness inconsistent formula.
2. By means of truth-tables, show whether [(A ⊃ B) ⊃ B] ⊃ (˥ A ⊃ B) is a tautology in S; show whether [(A → B) → B] → (˥ A → B) is a tautology in NS.

It is interesting to look at the argument forms of S, to see how many of them remain correct argument forms in NS. Such a comparison should give us some idea how S and NS are related. In fact, it turns out that S is an extension of NS. That is, all correct forms of NS are also correct in S. But there are some forms we found to be correct in S that are not correct argument forms in NS. Let us go through our list of logically interesting argument forms of S (from pages 187 and 188).

What about these two forms?

A ⊃ B	*modus*	A ⊃ B	*modus*
A	*ponens*	˥ B	*tollens*
∴ B		∴ ˥ A	

These forms are correct in S. But are they correct in NS?

A → B		A → B	
A		˥ B	
∴ B		∴ ˥ A	

Let us do a truth-table for *modus ponens*:

A	B	A R B	A → B
T	T	T	T
T	F	T	F
F	T	T	T
F	F	T	T
T	T	F	F
T	F	F	F
F	T	F	F
F	F	F	F

To test for correctness, we look to see if there is a row where both premisses, A and A → B, are true, but where the conclusion, B, is false. There is no such row; thus the argument form is deductively correct. A similar table will show that *modus tollens* is correct in NS.

However, when we come to the next argument form – "A, therefore A ∨ B" – we see that the form is correct in S, but is not correct in NS.

	A	B	A R B	A ∨ B
(1)	T	T	T	T
(2)	T	F	T	T
(3)	F	T	T	T
(4)	F	F	T	F
(5)	T	T	F	F
(6)	T	F	F	F
(7)	F	T	F	F
(8)	F	F	F	F

Rows (5) and (6) indicate that there are instances where the premiss A is true, but the conclusion, A ∨ B, is false. It is clear why this is so. The relatedness disjunction, A ∨ B, requires that A and B be related. So from the premiss that A is true, we cannot infer, always, that A ∨ B is true, for any B you like. A and B might be unrelated. In NS, the argument "Bananas are yellow, therefore either bananas are yellow or Scarborough is a suburb of Toronto" is not correct.

On the other hand, a disjunctive syllogism of a certain form is always correct in NS:

A ∨ B
⌐ A
──────
∴ B

If we have A ∨ B as a premiss, then we know that A must be related to B. Consequently, from ⌐ A we can always deductively infer that B is true.

Exercise

By means of truth-tables, show whether the following argument forms of NS are correct.

(a) A ∧ B
    ─────
     ∴ A

(b) A
    ⌐ B
    ──────────
    ∴ ⌐ (A → B)

(c) $A \rightarrow B$

$$\therefore \neg (A \wedge \neg B)$$

(d) $A \rightarrow B$
$A \rightarrow \neg B$

$$\therefore \neg A$$

(e) $\neg (A \rightarrow B)$

$$\therefore (B \rightarrow A)$$

Next, let us look at the logically equivalent forms of S. We see that certain laws, like (2a), (2b), and (10), still hold; others, like (1a), (1b), and (10), fail when we apply them to NS. So, again, S is stronger than NS. Every tautologous formula of NS is a tautologous formula of S. But some formulas that are logical equivalences in S fail as logical equivalences in NS.

In summary, we see that NS is as systematic and orderly a system as S, and shares most of its advantages. But in NS, the constants are defined differently, with the result that NS has many differences in its logically interesting properties. In particular, many argument forms that are ruled correct in S are not correct in NS.

Why should there be these differences? System NS is designed to do a different job than S. For in S, truth-values are the be-all and end-all of the way the constants are defined. In NS, however, the factor of relatedness of subject-matters is taken into account. So many arguments that are correct in S must fail in NS because they lack the appropriate relationships of subject-matters.

System S is open to *ignoratio elenchi* because it ignores the whole issue of subject-matters. But NS takes the relatedness of subject-matters into account, and therefore represents a different standard of correctness. Truth-table tests for correctness in NS are somewhat more complicated. And, to be sure, S is the stronger theory. But NS has a greater capacity to deal with the issue of subject-matters. Clearly, then, NS is a very useful theory if we are interested in the fallacy of *ignoratio elenchi*. But each system has its advantages and limitations.

6. Entailment Again

Each of our two formal deductive systems evaluates deductive correctness of arguments differently, although the two systems share many correct argument forms. It is all a matter of how we define the constants of the system; our definitions, in turn, depend on which features of an argument we want to include or ignore. NS takes into account the relatedness of subject-matters. Thus NS embodies a notion of argument whereby the *relevance* of a refutation is thought to be a significant factor in evaluating the correctness of an argument. S, on the other hand, is a simpler theory; it simply overlooks the whole question of relatedness of subject-matters.

Indeed we might say that S embodies a view of argument that presupposes that *all* propositions in an argument are related to each other. In other words, if we confine an argument to propositions that are all definitely related to each other in some topic, then we don't need to worry about relatedness at all, and we can use system S. But if there are some doubts about the issue of relatedness of subject-matters, then we should use NS, in order to account for the variation of subject-matters.

If we look at the arguments ruled to be correct in S, we can see that S does indeed seem to embody the view that relatedness does not matter. Consider these arguments, all of them correct in S:

$$(1) \quad A$$
$$\therefore A \lor B$$

$$(2) \quad A$$
$$\therefore B \supset A$$

$$(3) \quad \lnot A$$
$$\therefore A \supset B$$

These arguments have forms:

(1) A
Therefore, A or B.

(2) A
Therefore, if B then A.

(3) Not-A
Therefore, if A then B.

As substitution examples of each form, we could have the following arguments:

(1) Bananas are yellow.
Therefore, bananas are yellow or Joe Clark is Prime Minister.
(2) Bananas are yellow.
Therefore, if Joe Clark is Prime Minister bananas are yellow.
(3) Bananas are not yellow.
Therefore, if bananas are yellow Joe Clark is Prime Minister.

In S, as we saw, it is a law that a deductively correct argument must never take us from true premises to a false conclusion. And none of these arguments does that. For example, if it is really true that bananas are yellow, then at least one of the following is true: either bananas are yellow or Joe Clark is Prime Minister. As far as the standards of S are concerned, this argument must be ruled perfectly correct.

What is wrong with it as an argument? The two constituent propositions seem wildly unrelated. For that very reason, the argument cannot be ruled correct in NS. If subject-matters are to be taken into account, then such an argument must be evaluated as incorrect, an *ignoratio elenchi* fallacy.

In this chapter we have tried to bring out the advantages of the systematicity of formal deductive logic. Yet at the same time, the formal approach does not have to be locked into rigidity by cleaving to one unalterable single structure. Rather, its power lies in the flexibility that can be gained by extending the organization of classical, truth-functional logic to cope with various aspects of argument, while still retaining the precision of its formal methods.[2]

Notes

1   The important connection between an argument form's deductive correctness and a conditional formula's truth-functional tautologousness is established by what logicians call "the Deduction Metatheorem". Interested readers might consult John Woods, *Proof and Truth* (Toronto: Peter Martin Associates, 1974), pages 110; 169–171; and Howard DeLong, *A Profile of Mathematical Logic* (Reading, Massachusetts: Addison-Wesley, 1970), page 134ff.

2   For further study, see I.M. Copi, *Symbolic Logic*, 5th ed. (New York: MacMillan, 1979). Two articles explain more about the motivation and structure of relatedness logic: Richard L. Epstein, "Relatedness and Implication," *Philosophical Studies* 36 (1979): 137–173; Douglas N. Walton, "Philosophical Basis of Relatedness Logic," *Philosophical Studies* 36 (1979): 115–136.

Chapter Ten
Economic Reasoning

"Economics", said Lord Carlyle, "is the dismal science."[1] A tough-seeming verdict, to be sure, but today many people mistake the import of Carlyle's remark. They think Carlyle meant that economic reasoning was *characteristically* of poor quality, that economic reasoning tended to involve too much bad logic or bad methodology; or, more simply, that economics was not properly scientific and therefore too often gave rise to reasoning that was incorrect. It is true that economic reasoning can go wrong, and that it can give rise to fallacies. But Carlyle was not concerned with fallacies. His remark was about a doctrine put forward by Thomas B. Malthus, a nineteenth-century British economist. Malthus observed that population growth was exponential (that is, geometric) whereas growth in the food-supply was arithmetical. It could be predicted, then, that exponentially growing populations would be degraded by the slower growth of the food supply, and the supply of other necessities of life. And Malthus himself witnessed such degradation in England. People were starving. And they were exposed to various forms of depravity — the consumption of drugs and alcohol was out of control, infant mortality rates were at a historical high, and civil and moral corruption abounded. A dismal story, if ever there were one. And the story prompted Carlyle's remark, and the words "the dismal science." The remark's implication has survived the development of economics even to the present day. The American economist Kenneth Boulding characterizes two of Malthus' most serious findings as the "Dismal Theorem" and the "Utterly Dismal Theorem". According to the Dismal Theorem, if the sole effective check on the growth of a population is misery, then the population will expand until its misery forces it to stop. And according to the Utterly Dismal Theorem, technological improvements permit populations to expand and thus provide for an increase in the number of those who will be miserable.[2]

The great Romantic poet, Samuel Taylor Coleridge, was incensed by Malthus' economic principle. Coleridge wrote: .

> I do not believe that all the heresies and sects and fashions, which the ignorance and the wickedness of man have ever given birth to, were altogether so disgraceful to man as a Christian, a philosopher, a statesman or a citizen, as this abominable tenet.[3]

Exercise

Scrutinize Coleridge's statement. Does it commit any fallacies or dialectical indiscretions?

1. The Fallacies of Composition and Division

Clearly, Carlyle was not saying that Malthusian reasoning is incorrect, but only that its conclusions are rather grim. Still, owing to the obvious importance of economic reasoning for personal, domestic, regional, national and international affairs, it is fair to ask whether there are fallacies to which economic reasoning is *particularly liable*. The answer is yes.

Economics can be divided into two components, *microeconomics* and *macroeconomics*. Microeconomics is the economics of particular *markets*, such as the auto industry or the toothpaste industry. Macroeconomics is the economics of *systems of markets*, of the whole network of markets that makes up an *entire economy*, such as the economy of Japan or the economy of the USA. Roughly speaking, markets (microeconomics) are the *component parts* of whole economic systems (macroeconomics). Thus, microeconomics stands to macroeconomics somewhat as parts stand to wholes.

One traditional fallacy, called the Fallacy of Composition, involves confusion: the reasoning that applies to parts is confused with the reasoning that applies to wholes. Economics affords some interesting and instructive illustrations of this fallacy.

The Fallacy of Composition: It is often mistakenly assumed that what is true for the parts of a system is true for the system as a whole. If you stand up at a football game, you can see better, but if everybody stands up nobody can see better.

In economics, if you, as an individual, decide to save more out of your income, you will increase your wealth. But if everyone in the nation tries to save more out of income, this may reduce national wealth – by reducing, in succession, sales, the production of goods, the incomes of producers and their employees, and ultimately national saving and investment.

If you, as an individual, are able to raise your prices, that may be a good thing for your business. But if every business [in the national economy] does the same the obvious result will be inflation, a bad thing for the nation.

Balancing the budget so that outgo does not exceed income may be a sound rule for you and your family. But budget balancing does not always make sense for the national government; for the government to do so during a business slump when unemployment is rising would worsen the slump and increase unemployment.[4]

What can we conclude from these warnings? Leonard Silk provides an answer:

. . . when we shift from micro- to macroeconomics some key concepts change.[5]

In other words, *reasoning that may be valid for parts may be seriously inadequate for the whole*. To ignore this warning, whether in economics or in other branches of reasoning, is to risk committing the fallacy of composition.

It would be unfair and misleading to suggest that economists commit that fallacy more than other thinkers. But economics is one of the sciences that has two distinctive sub-branches, microeconomics and macroeconomics. The branches are related as part to whole, and therefore economics is *inherently liable* to misapplying reasoning that is valid for parts to wholes.[6]

Make no mistake about it: people are more than ready to confuse micro- and macroeconomic reasoning. Consider this exercise. Does it or doesn't it involve the fallacy of composition?

Exercise

Mr. Businessman: What would happen if I ran a deficit in my hardware business?! I'll tell you what would happen: I'd be bankrupt. I'd have no business left. Yet the government has run a deficit of several billion dollars and the national economy is running a huge balance of payments deficit. Face it, the whole country is out of business, and we're just too thick-headed to realize it!

Ms. Businesswoman: It is quite clear that when I cut my employees' wages last year, I could afford to hire some additional people. That brought unemployment in my town down a bit. So the best thing to do for unemployment right across the country is to have all employers cut wages.

Economic reasoning is prey to the fallacy of composition. And this fallacy crops up in other contexts as well. Here are some examples:

(1) Two aspirins are effective for a headache, so the whole bottle should really do the trick.
(2) Each part of Mohammed Ali weighs less than fifty pounds, so Ali isn't really a heavyweight.
(3) All the players who play for the San Diego *Chargers* are good, so the *Chargers* are a good team.

Mind you, not every instance of reasoning from what is true of parts to what is true of the whole involves the fallacy of composition. Here are some examples of perfectly sound reasoning:

(1) All the parts of the machine are made of iron, so the machine is made of iron.
(2) The bikini top and the bikini bottom are made of cotton. Hence, the bikini is made of cotton.

And sometimes it is very difficult to determine, in a general way, whether reasoning from part to whole is correct. For example, if Sue kicks Bill's shin (and so kicks a part of Bill) she kicks Bill. And this might lead us to believe that

(3) Anyone who kicks a part of Bill kicks Bill.

But what if Sue kicks the end of a long strand of Bill's hair? Say that Bill, who is a bit of an eccentric, has grown this strand to a length of sixty inches. Did Sue kick Bill?

The Fallacy of Division   The other side of the fallacy of composition is the *fallacy of division*. If composition is the fallacy of reasoning from parts to the whole, then division is the fallacy of reasoning from the whole to its parts. Since division is also a fallacy of parts and wholes, it, too, can crop up in economic reasoning. Here are some quite obvious examples of the fallacy of division:

(1) The United States has, in 1980, a Democratic Congress; therefore Congressman John Anderson is Democrat.
(2) This machine is heavy, so all its parts are heavy.
(3) Boston has one of the most highly educated populations in North America; consequently Bruiser Savage, who lives in Boston, is one of the most highly educated persons in North America.

The Fallacy of Over-Rating   Consider the following argument:

(i) If it is more likely than not, that these premises are true (i.e. that they are both true), it is more likely than not that the conclusion is true;

(ii) but it is more likely than not that the premisses are true (i.e. that *each* of them is so);

(iii) *therefore*, it is more likely than not that the conclusion is true.[7]

It can be shown that this argument contains a fallacy of division. To demonstrate, we introduce the concept of *befortification*. Very simply, p befortifies q if and only if $Pr\,(q, p) > \frac{1}{2}$. That is, p befortifies q where the conditional probability, q given p, is greater than one-half. Moreover, a *simple* proposition, p, is *befortified as it stands*, if its probability is greater than one-half. We represent the form of our sample argument as follows. Let P_1 and P_2 be two independent premisses; let C be a conclusion. The sample argument can be restated:

(ia) If $(P_1 \wedge P_2)$ is befortified then C is befortified.
(iia) P_1 is befortified and P_2 is befortified.
(iiia) Therefore, C is befortified.

The invalidity of the argument shows up very nicely in elementary probability theory:

(ib) If $Pr\,(P_1 \wedge P_2) > \frac{1}{2}$ then $Pr\,(C)$ $\frac{1}{2}$.
(iib) $Pr\,(P_1) > \frac{1}{2} \wedge Pr\,(P_2) > \frac{1}{2}$.
(iiib) Therefore, $Pr\,(C) > \frac{1}{2}$.

Exercise

Verify the incorrectness of argument (b).

What is true for the whole group of premisses – namely, the group P_1, P_2 – is not necessarily true for each premiss individually. Fallacy of division!

2. Analytical Remarks on Composition and Division

As we have said, the fallacies of composition and division involve reasoning incorrectly from parts to whole and from whole to parts. But not all reasoning from parts to whole or whole to parts is incorrect or fallacious. Can we determine, in a suitably general way, *when* such reasoning is fallacious?

To deal with this question, we introduce two simple definitions. Let W be a whole (or aggregate); let F be some property or attribute; let $P_1, \ldots, P_n = P_i$ be the parts of the whole (or aggregate) W. Then:

(C): Property F is *compositionally hereditary* with regard to aggregate W if and only if, every P_i has property F, when W itself likewise has property F.
(Example: if all the parts of the machine are metal, then the machine is metal.)

(D): Property F is *divisionally hereditary* with regard to aggregate W if and only if, W has property F, when so, too, do all its parts, P_i, have property F.
(Example: if this team is the Canadian Olympic Team, then its members are all Canadians.)

Perhaps the most important analytic fact about the fallacy of composition and division is this:

> For very many wholes or aggregates, some properties are compositionally hereditary and some are divisionally hereditary and some are neither.

Also:

(L1): For any aggregate, W, there is some property, F, that is *not* compositionally hereditary.

(L2): For any aggregate, W, there is some property, F, that is *not* divisionally hereditary.

It is easy to verify (L1). Suppose that we have an aggregate, A, all the elements of which are *non*-aggregates. Then A is not compositionally hereditary as regards the property of being an aggregate!

Exercise

Construct an example that verifies (L2).

There are two further laws that we can cite.

(L3): If F is compositionally hereditary as regards W, it does *not* always follow that F is divisionally hereditary as regards W.

(L4): If F is divisionally hereditary as regards W, it does *not* always follow that F is compositionally hereditary as regards W.

Here is an example that verifies (L3): let F be the property of weighing more than two pounds. Clearly this property is *compositionally* hereditary with respect to a machine, but it is *not divisionally* hereditary. That is, if all the parts of a machine weigh more than two pounds, so does the machine. But although the machine might weigh more than two pounds, all its parts might not.

Exercise

Construct an example that verifies (L4).

The Logical Form of Composition and Division.

It is evident that composition and division have some elementary logical forms:

> *Composition*
> All the P_i have F
> Therefore *W* has F
>
> *Division*
> *W* has F
> Therefore all the P_i have *F*

And it is also clear, because of (L1), (L2), (L3), and (L4), that *when* composition is a fallacy it is *not* so solely by virtue of its elementary logical form. And *when* division is a fallacy it is *not* so solely by virtue of its elementary logical form. Even *when* an argument, of which the elementary logical form is that of composition, *is a fallacy*, it does *not* follow that the re-arrangement of that self-same argument, of which the elementary logical form is that of division, is likewise a fallacy.

So we have not yet attained much generality about when composition and division are fallacies, and we have not yet said what, beyond an argument's elementary logical form, *makes* a composition argument or a division argument fallacious.[8] We can say this much here: Whether argument in the form

> All the P_i have F
> Therefore *W* has F

or in the form

> *W* has F
> Therefore all the P_i have F

is fallacious depends upon

(a) the precise nature of the property F and the precise nature of the aggregate *W*; and
(b) a suitably powerful general *theory* of the part-aggregate relationship.

Basically, the precise kind of property and aggregate involved in the argument makes a real difference as to whether the argument is fallacious; it is more than a matter of elementary logical form. And the analysis of part-whole relationship is also relevant to our question. This much we do know. If you look to ordinary mathematics, *set theory*, for your analysis of the part-whole relation, your account of compositional and divisional fallaciousness will quickly turn into a ludicrous disaster.[9] The moral: for an adequate analysis of composition and division, *more than classical logic and set theory is required.*

3. Economic Reasoning

It would be quite wrong to leave the impression that the most distinctive logical feature of economic reasoning is that it may commit the fallacy of composition and division. Another of its distinctive features is its concern with *decision-making*. Leonard Silk characterizes the "problem-solving" function of economic reasoning:

1. *Define the problem.* For example, is the problem to secure tactical or strategic advantage? According to some military analysts, the Dieppe Raid during World War II was a *tactical* disaster, involving terrible losses to the Canadian landing troops, but an excellent *strategic* victory in that it taught the Allies how to conduct the later Normandy Invasion effectively. So, if the problem was to secure tactical advantage, then the problem-solving reasoning that led to Dieppe was catastrophically mistaken; but, if the problem was to secure strategic advantage, this problem-solving reasoning may have been quite successful.

2. *Identify the goals and values* that you want a solution to achieve. For example, you are a member of a trade union, and your group is very worried about the corrosive effects of inflation. As the saying goes, "You want to keep ahead of inflation". In your settlement with management it is agreed that your wages will increase by exactly the percentage by which the cost of living index exceeds 5%. In fact the cost of living index stands at 6%, and you are rejoicing in the good news of a 1% salary *increase*. But notice, it is *not* what you wanted or valued, and it is *not* what you aimed for, for your purchasing power *declined* 5%. You did *not* keep ahead of inflation; inflation beat you by 4 points net!

3. *List alternative courses of action.* With some exceptions, *single-solution* problem-solving is asking for trouble. Suppose that you believe that your weekly paycheck is about $15.00 short of permitting you to make ends meet. Even though you know that your industry is economically very weak and uncompetitive, and that the regional economy in which your industry is set is also in a bad way, you exercise your influence as local union president, and you call a strike. If your goal is to augment *your* income by $15.00 a week, almost certainly you should have at least considered some alternatives: moonlighting, qualifications upgrading, changing jobs, etc.

4. *Weigh each alternative.* It is not enough just to *consider* the alternatives; they must also be analyzed and weighed. Comparisons should be made and the various outcomes should be predicted.

5. *Choose the alternative that seems to solve your problem best, and then act on it.*

6. *If possible, after you have acted, go back and check the results.*[10]

Dialectic and Decision-making   What is required to make a deliberate, reasoned decision? A decision, say, that conforms with Silk's six general precepts? You must engage someone (often *yourself!*) in a kind of *dialectical argument*. Recall, from Chapter Six, that a dialectical argument is one in which, by question and answer, a participant attempts to refute a thesis. The rules of testing are carefully arranged to preserve fairness and to enhance the chances of arriving at the correct answer. Consequently, if the original thesis is *refuted*, then the correct answer is the negation of that thesis. If a refutation cannot be made, the original thesis has not been proved, but that thesis has attained some rational support. And the supporter of the thesis can continue to support it.

Let us suppose that the *original thesis* is "Decision X should be chosen". Suppose, further, that questioner and answerer are one and the same person. This is an *allowable extension* of Rule of Dialectic 1 (see page 111), provided that the characteristic *rôles* of questioner and answerer are preserved and honoured. This extension of Rule 1 expressly provides that dialectic may be transacted in *solitaire* versions.

Imagine that our dialectician, Bill, puts to himself a question which, as answerer, he himself cannot answer. According to the strict rules of dialectic, the argument must then stop, for, by Rule 5, the questioner is expected to help overcome the ignorance of the answerer. But here Bill is both questioner and answerer. However it is an *allowable extension* of Rule 5 for Bill, as answerer, to solicit independent advice about the question, to which he does not know the answer – provided that the information he solicits is purely factual and objective, and provided it comes from an independent source with no stake in the outcome of the dialectical argument in progress, and provided, finally, that, *in his persona as questioner*, Bill finds this method of research acceptable. "Research" is the key word. Our extension of Rule 5 permits the consultation of a factual *research* source, such as an encyclopaedia, or a file clerk in the Hall of Records, or a University Registrar.

Clearly, if the original thesis is refuted, then the correct decision is *not* to do X. At this stage, it is appropriate to move to the next decisional possibility and to test it dialectically. If it, too, is refuted, Bill can move on, testing all the decisional alternatives. Thus, in our extension of our dialectical rules, we have an additional principle:

(9)   If a unique decision is desired, cite, in order, each decision in the range of decision alternatives as a thesis, and test for its correctness dialectically.

We can formulate the process:

(a)   if, in a given range of decision alternatives $(D_1, \ldots, D_r)$, exactly one, D_i, is left unrefuted, then it is generally reasonable to choose D_i;

(b)   if all alternatives are refuted, then it is generally reasonable to do nothing;

(c) if more than one decision alternative is left unrefuted, then formulate additional *comparative* theses, that is, theses in the form "Decision D_i is more advantegeous than decision D_j", and test dialectically;

(d) if, in the end, no *comparative* thesis emerges as the dialectical winner, then in general, "Do nothing" is a rational choice. It is crucial to understand, however, that there exist at least two serious exceptions to "Do nothing".

(A) *The forced option.* You are standing in a narrow street without sidewalks and a large truck, whose driver has lost control, is quickly bearing down on you. Your options are to spring through the open door on your left (the only route of escape) or to stay put and have an early death. You've been very depressed recently, and indecisive, and now you really can't make up your mind. So you reach this decision: "Do nothing!" But that is equivalent to thinking, "Stay and get killed". You are in a *forced option*, as the philosopher William James called it. A forced option is one in which "doing nothing" is really "doing something", that is, your postponement is equivalent to one of the decisions you wished to avoid making.

(B) "Do nothing" is rather conservative advice; it is designed to minimize risk and to promote informed decision-making. But some issues, certain emergencies, for example, require that even when you don't know what to do, an uninformed try is better than doing nothing whatsoever.

The theory of decision-making, like the logic of dialectic of which it is a part, is designed to articulate various *canons of rationality* in a useful way. It does not promise to be a miracle problem-solver.

Decision Theory   Modern decision theory comprehends four main components:[11]

> NORMATIVE
> THEORY
> *(Normative theories*
> *involve judgment*
> *of value)*
>
> DESCRIPTIVE
> THEORY
> *(Descriptive theories*
> *preclude judgments*
> *of value)*
>
> INDIVIDUAL
> DECISIONS
> Classical economics
> Statistical decision theory
> Moral philosophy
> Experimental decision studies
> Learning theory
> Surveys of voting behaviour

GROUP
DECISIONS
Game theory
Welfare economics
Political theory
Social psychology
Political science

In citing these various branches, we expose much more of decision theory than we will have time for here; nevertheless, it will be of some use to have a rough idea of the scope of contemporary investigations into decision-making. That said, here is the leading theme of decision theory:

a) a person or group of persons is exposed to a number of action-alternatives;
b) they do not possess complete information about the current state of affairs;
c) they do not possess complete information about the results and implications of each possible action.

Consequently, the *decisional problem* is

(1) to select an action that, relative to the available information, is *rational* and, if possible, *optimal* (the best among the available rational strategies);
(2) to make this selection, in conformity with *exact criteria* of what is rational and optimal.

Normally, decision theory is taken also to include *minimax theory*. The central idea of minimax theory is that estimates and choices should be made so as to "minimize the maximum possible expectation of error".[12] Minimax theory belongs mainly to *statistical decision theory*, and thus is a part of normative decision theory. We will have more to say of minimax theory later in this chapter.

A vexing problem for the normative theory of individual decision-making lies with the concept of *rationality*. A major concern for this branch of the theory has been to shed strong theoretical light on the very idea of rationality. The trouble is that contemporary work in decision theory tends to show, rather convincingly, that there seems *not* to be any simple and consistent set of principles that admit of exact statement and that correspond, in a natural and intuitively satisfying way, to the general idea of rationality. This situation, which is theoretically very frustrating, is similar to a situation in the foundations of mathematics; it has become rather clear that we don't really know, in a theoretically well-articulated way, just what the *mathematics of sets* is! Both in decision theory and in the foundations of mathematics, we see that definitions of key concepts, such as rationality or set, are definitions that seem very plausible, but actually *lead to paradox*. In this respect it may be helpful to recall how Tarski and others showed that from the innocent-seeming idea of *a semantically closed language*, horrendous contradictions ensue (see Chapter Eight).

Rationality and Utility   In a historically important passage, Jeremy Bentham postulated an interesting connection between rationality and *utility*. "By utility", Bentham wrote,

> is meant that property in any object whereby it tends to produce benefit, advantage, pleasure, good or happiness (all this in the present case comes to the same thing), or (what comes again to the same thing) to prevent the happening of mischief, pain, evil, or unhappiness to the party whose interest is considered: if that party be the community in general, then the happiness of the community, if a particular individual, then the happiness of that individual.[13]

For Bentham, then, rationality consists in *maximizing utility*, and this idea became a dominant theme of *classical economics*, as it is called. However, classical economists have had enormous difficulty in formulating exact measures of utility.

The *problem of measuring utility* was partly solved in the beginning of this century by the Italian economist Vilfredo Pareto. Pareto demonstrated that classical economics could get on quite nicely provided it could enable a person to know which of any *two* decisions had consequences of greater utility. Pareto's work depended upon a rather dubious assumption, however; namely, that a person who is faced with a choice among alternatives *is not uncertain about the consequences of these alternatives*. If, on the other hand, uncertainty of such consequences is admitted, it would not be possible to order the possibilities as first-best, next-best, and so on, and the rational thing to do would not admit of a fully determinate specification. Consequently, if decision theory does not grant this implausible assumption, it must develop its theory of decision in such a way that rational decisions are not necessarily optimal decisions, or first-best decisions.

Uncertainty of Outcome   It is obvious that rational decisions can be and are being taken even where there is uncertainty about the consequences of the available alternatives. But just how are rational decisions arrived at when one is uncertain of the consequence? A very important part of the answer comes from an eighteenth-century mathematician, Daniel Bernouilli.

Imagine a person who is faced with several decision possibilities. These decisions may have various consequences. Furthermore, it will not generally be true that the act of deciding to adopt a given possibility will trigger, all by itself, the consequences of that decision. Consequences also usually depend on, or are influenced by, the entire state of affairs that obtains when the given decision is taken. Thus rational choice normally has two parts:

213

(1) the decider must associate utilities with the various possible outcomes;

(2) the decider must consult his beliefs about the present state of affairs.

As for how the decision is rationally made, let us turn to a concrete example. Suppose that Bill must decide whether to attend a play in uncertain weather. So we will, for simplicity, imagine that the present state of affairs encompasses the possibility of its snowing, S, and the possibility of its not snowing, not-S. Suppose, further, that the possible consequences are:

C_1: going to the theatre and not being snowed on.
C_2: staying home.
C_3: going to the theatre and being snowed on.

Bill's two decision possibilities are:

D1: going to the theatre
D2: not going to the theatre.

Imagine, further, that Bill believes it is more probable that it will not snow than that it will snow. Accordingly, he assigns to each possibility its own real number; $4/10$ to possibility S, and $6/10$ to possibility not-S. In this way, he gives numerical expression to the "subjective probabilities" of S and not-S. For Bill, the subjective probability of its snowing or not snowing is the numerical expression that he gives to his degree of confidence that it will or will not snow. In our scenario, his degree of confidence that it will snow is expressed as the subjective probability $4/10$; his degree of confidence that it won't is expressed as the subjective probability $6/10$. Let us also say that Bill prefers conclusion C_1 to conclusion C_2; he prefers C_2 to C_3.

It is clear that Bill's two decision possibilities confront him with three possible outcomes. If he chooses D1, then he gets what he most prefers, namely C_1, or what he least prefers, namely C_3. And if Bill selects D2, he is certain to achieve his middling preference, C_2.

Obviously Bill is not quite ready for a rational choice. He must now very carefully evaluate his preferences. This he can do by assigning precise numerical values to the three consequence possibilities. In this way he can express precisely the utility of C_1, C_2, and C_3. Accordingly, Bill makes these assignments:

C_1 is assigned 10
C_2 is assigned 5
C_3 is assigned -10

Bill now has what he needs to attempt to make a rational decision. He will estimate the *expected utility* of each decision possibility, D1 and D2. The expected utility of D1 we will call E(D1); of D2, E(D2). How, then, does Bill compute E(D1) and E(D2)?

(1) For E(D1), Bill calculates the product of the probability that it will

snow and the utility of his being snowed on; to this he adds the product of the probability that it won't snow and the utility of his going to the theatre without being snowed on. Thus:

$$E(D1) = (^4/_{10})(-10) + (^6/_{10})(10) = 2$$

(2) For E(D2), Bill takes the product of the probability of its snowing and the utility of staying home; to this he adds the product of the probability of its not snowing and the utility of his staying home. Thus:

$$E(D2) = (^4/_{10})(5) + (^6/_{10})(5) = 5$$

Consequently, according to the expected utilities approach, the rational thing for Bill to do is to make decision D2. Bill should stay home, for that is the way for Bill to maximize his expected utility.

4. Expected Utility, Sure-Thing and Minimax Principles

If you look at things the way Bentham did, you will tend to believe that utility is *pleasure*. (Remember that Bentham said benefit, advantage, good, happiness and *pleasure* "all . . . come to the same thing".) But surely it is not even plausible – or well-argued – to think that only those things which are to your advantage, or which confer a benefit on you, are your own pleasure! And surely (you say) a decision theory that rests on such a notion of utility and expected utility is severely implausible.

Strangely enough, the facts are to the contrary. Decisions can be computed quite satisfactorily no matter what you mean by the word "utility". Provided that your idea of utility includes the thought that a utility is something valuable, its precise interpretation does not matter.

But it is easy to see that this solution to the difficulty takes us out of the frying pan and into the fire. For if "utilities" can mean just about anything that is valuable, then utility theories of decision are seriously incomplete. They tell us little about what rationality really is, especially since values often conflict with one another.

Here is a maxim of rationality: "maximize your expected utility". The maxim doesn't tell us very much about rationality, and so gives rather incomplete advice. But it also asks too much of us. It expects us (as it expected Bill) to provide uniquely exact numerical expressions of our beliefs about the present state of affairs, and uniquely exact numerical expressions of our outcome-preferences. The plain fact is that very often a decision-maker will not have adequate information to assign exact probabilities to possible states of affairs or to assign exact preferences to possible outcomes. The principle of the maximization of expected utility is, therefore, *too demanding*.

Less Demanding Decision Principles: Sure-Thing and Minimax

Games theory and statistical decision theory are two branches of general decision theory that seek to formulate weaker decision principles (or decision principles that are weaker than the maximization of expected utility, principles that do *not* require the decision-maker to know the exact probability of each event relevant to the decision at hand).

One such weaker principle is the *sure-thing principle*. Imagine that our decision-maker, Bill, is faced with two decisions, and that for each possible existing state of affairs, the consequences of selecting the first decision over the second are *at least as desirable* as those of selecting the second decision over the first. Then the sure-thing principle asserts that Bill should *prefer* the first decision.[14]

What should we say about the sure-thing principle? It is a fine thing in theory, and it makes a contribution to the mathematical formulation of decision-making principles, but it has *very limited application* in situations of real-life decision-making. Recall what this principle requires of the decision-maker. Bill must know the consequences of *every* possible state of affairs relevant to his decision possibilities, and he must know, for each set of consequences, which, if decided upon, would be more or less desirable than the others. But it is obvious that, relative to all sorts of real-life decisional situations, Bill is required to know too much. The sure-thing principle often imposes unrealizable tasks.

Another interesting decision principle is the John von Neumann's *minimax principle*. The minimax principle is central to Game Theory.[15] Put at its simplest, Game Theory is the scientific study of *conflict of interest*. It deals with "conflict, collusion, and conciliation".[16] Imagine that Bill is faced with a number of decisions, and that he is able to determine the worst possible outcome. The worst outcome will be that of the *minimum* desirability. Bill must then select that decision for which the *worst* outcome is as *desirable as possible*; that is, Bill selects the decision that is the most desirable if its least desirable consequence occurs. Such a decision is called a *maximum decision*; it is a decision that minimizes the maximum loss or maximizes the minimal gain.

Zero-Sum Games To see how maximal decisions work, imagine yourself in a *two-person zero-sum game*. A zero-sum game is one in which the participants are in a conflict of interest, and in which any winning strategy implies that the game must also have losers. Let us say you and Bill are playing a two-person zero-sum game. If you win, then Bill loses; moreover, if you win, you win *precisely to the same degree* that Bill loses. Thus if the utility attaching to your win is expressed numerically as $^6/_{10}$, then the *disutility* attaching to Bill's loss must, when summed with your $^6/_{10}$, yield zero. Hence the disutility of Bill's loss is $-$ $^6/_{10}$.

$$(^6/_{10}) + (- \ ^6/_{10}) = 0$$

According to von Neumann, in many two-person zero-sum games, each player has a *minimax strategy*; that is, a strategy designed to minimize his maximal loss. Furthermore, says von Neumann, if both participants are rational, the best possible strategy for each is to select his own minimax strategy.

Notice how conservative are minimax strategies. Their object is to make the best of a bad lot, and they condition the participants to play it safe and to accept an outcome with a low "pay-off". There is no doubt, as we said earlier, that in certain real-life situations, the most rational course *is* conservative, and to make the best of a bad lot is exactly right. But it is also true that, for any rational person, some decision strategies demand high pay-off even at high risk. In life-and-death situations, for example, the surgeon cannot always elect to make the best of a bad lot by letting the patient expire, say, unoperated upon but comfortably medicated.

5. Group Decisions and New Welfare Economics

As we have seen, a leading idea of classical economics is that of the greatest pleasure of the greatest number, or the maximizing of utilities. Bentham's interest in maximization had to do primarily with mechanisms for the distribution of economic benefits. But many thinkers have come to see that such mechanisms are interesting even when they deliver benefits that are non-economic. In fact, a major concern of *welfare economics* is to investigate schemes for distributing any kind of good, whether it be economic benefits (such as *income*) or non-economic benefits (such as *social justice*). Bentham's insight, therefore, can be turned into a generalization, a problem for group decision theory: *how to distribute benefits over groups with maximal pay-off?*

We have seen that *individual* decision theory has not been very successful in formulating intuitively satisfying criteria of rationality. The same can be said for *group* decision theory. Interesting work in this connection has been done by Kenneth J. Arrow.[17] Arrow wishes to know the extent to which social decision can be equitable. Consider a society of individuals. This society will, at any given time, have available to it a number of different states of affairs, or *social states*. Suppose that each individual member of the society is able to rank or order these various social states according to his or her own individual preference. How, then, do we arrive at a just *social* preference-ordering from these *individual* preference-orderings? Or, in concrete terms, what is the preferred social state for the *country*, given the preferences of its citizens? And can the country's preference be arrived at fairly and equitably?

Perhaps social preferences might be ranked by *majority decision*. One social state, X, would be preferred to another social state, Y, if a majority of individual citizens prefer X to Y. However, it is far from clear whether majority decisions are always equitable, as can be seen by an examination of the so-called Voters' Paradox.[18] Imagine that there are three issues to be put to the vote: I_1, I_2, and I_3. For simplicity, let us suppose that there are just *three* voters voting on these issues.

> Assume that the first voter prefers I_1 to I_2, and I_2 to I_3.
> Assume that the second voter prefers I_2 to I_3, and I_3 to I_1.
> Assume that the third voter prefers I_3 to I_1, and I_1 to I_2.

The vote will be taken in pairs. That is, if the first vote counted is between I_1 and I_2, then I_1 will be selected. If the first vote is between I_1 and I_3, then I_3 will be selected; if the first vote is between I_2 and I_3, then I_2 will be chosen.

Exercise

Verify these assertions.

Sometimes the outcome of a vote will depend on the arbitrary selection of which issues are voted on *first*! And this squares with actual voting experience, in which the order of voting on and amending Congressional or Parliamentary bills can drastically influence the outcome. In general, then, the device of the majority vote is not always a rational or just mechanism for the fixing of social preference.

Pareto's Welfare Economics How are we to obtain "the maximum of well-being for a collectivity"? Here is Pareto's own answer:

> Consider any particular position and suppose that a very small move is made [from it] . . . [Then if] the well-being of all the individuals is increased, it is evident that the new position is more advantageous for each of them; vice versa it is less so if the well-being of all the individuals is diminished. The well-being of some may remain the same without these conclusions being affected. But if, on the other hand, the small move increases the well-being of certain individuals, and diminishes that of others, it can no longer be said that it is advantageous to the community as a whole to make such a move.[19]

Here we meet the idea of the so-called *Pareto improvement*. A Pareto improvement, on a given distribution of benefits by an alternative distribution of those benefits, is one in which every individual does *no worse* than he does on the initial distribution. Consider a given distribution of benefits, B; and consider a set of alternative distributions to B. Now if *none* of these other distributions makes a Pareto improvement on the initial distribution B, then B is called a *Pareto optimal* distribution.

According to Pareto, the rational and just strategies of social decision-making are those that *promote Pareto optimality*. But there are problems with this approach. All too often, in real life, *no* alternative distributions are Pareto improvements, for each requires that at least one person would have to surrender at least some slight advantage for the advantage of the others.

Another difficulty is that, in the real-life practicalities of social and economic life, there is frequently no *best* alternative strategy; that is, there is no absolutely dominating Pareto improvement. And there is nothing in Pareto's theory to guide us in the preference ranking of, and the preferential choice from among, *suboptimal alternatives*.

The New Welfare Economics Over the decades, welfare economists attempted to develop their theories so that such problems might be dealt with. Prominent among later developments has been the school of the "New Welfare Economics", which originated in a manifesto by Nicholas Kaldor.

Suppose that a society is faced with deciding how to distribute *income*. Kaldor proposed the important idea that the distribution of income was *not* an economic question; instead, Kaldor said, it was an *ethical* question. Accordingly, in all cases

> where a certain policy leads to an increase in physical productivity, and thus of aggregate real income, the economists' case for the policy is quite unaffected by the question of the comparability of individual satisfactions; since in all cases it is *possible* to make everybody better off than before, or at any rate to make some people better off without making anybody worse off. There is no need for the economist to prove – as indeed he never could prove – that as a result of the adoption of a certain measure nobody in the community is going to suffer. In order to establish his case, it is quite sufficient for him to show that even if all those who suffer as a result are *fully compensated*, for their loss, the rest of the community will still be better off than before. Whether [a given group] should in fact be given compensation or not is a political question on which the economist, *qua* economist, *could hardly pronounce an opinion*. [Italics added][20]

That is, economists should deal with increasing production and should not trouble themselves with the distribution of the increased income. But what justifies this rather cynical doctrine? Kaldor thought that it could be justified as follows: Whenever the total quantity of goods and services is increased, then

(i)   it is always possible for those who benefit from this growth in production to persuade the losers to accept the new situation, because the losers can always be compensated or rewarded for adjusting to the new scheme of production rather than sticking with the old situation;

(ii)   thus, from the point of view of the economist, the new situation is preferable to the old.

This is *the Kaldor Compensation Principle*. It rests on the outright misconception that beneficiaries are likely to compensate losers in a new economic situation! Moreover, there is no good reason for thinking that the Kaldor Compensation is even possible. In general, the product of an economy is *not* redistributable. Certain *theoretically possible* redistributions could be made only at the expense of the malfunction of the whole economy.

Still, for all its difficulties, the Kaldor Principle has significantly influenced modern Anglo-American economics. The general trend of welfare economics has been to avoid normative principles or judgments of value – so the trend is not very effective in *evaluating* distributions.

Reprise   It is useful to classify decision theory as an extension of dialectic; thus it is helpful to think that economic reasoning is dialectical in character.

(1)   Moreover, since it is quite evident that the logic of decision theory is far from complete, it follows that the logic of dialectic is also far from complete.

(2)   It is also worth mentioning that, apart from its general susceptibility to dialectical lapses, economic reasoning is subject to a number of fallacies and paradoxes:
   – the fallacy of composition and division
   – the fallacy of over-rating
   – the voters' paradox

6.   The Prisoners' Dilemma, and What It Suggests about Rationality

We said at the very beginning that logic is a science of reasoning. When correctly used, reasoning is a powerful and effective manifestation of *rationality*; when badly done, it is a disturbing manifestation of imperfect or damaged or degraded rationality. We have seen that one of the negative results of decision theory is that the concept of rationality seems not to be understood in a theoretically satisfying way. This sounds a cautionary note for logic, of course; but it is even worse news for economics. In classical economics, especially, one finds a quite ready identification of rationality with *self-interest*. However, one especially interesting outcome of Game Theory casts serious light on this identification. We turn now to the *Prisoners' Dilemma*.

Bill and Sue are prisoners, held incommunicado. Each is charged with being an accomplice to a serious crime. However, a conviction cannot be secured unless Bill and Sue can be induced to incriminate each other. So, if Bill and Sue each incriminate the other (that is, if there is mutual incrimination) each will divide the resulting prison term of twenty years; that is, each will be jailed for ten years.

However, if Bill "talks" and Sue keeps silent, Sue will be convicted; she will serve the whole twenty years. If Sue "talks" and Bill keeps quiet, Bill will serve the entire term.

If both keep silent – that is, if there is no incrimination – each will be convicted on a much lesser charge, and each will serve a prison term of one year.

What course should each follow? Here we have a two-person game, which we can represent in a table below. This table is also called a "pay-off matrix". The numbers in the matrix are numerical expressions of gains (or pay-offs) and losses (or negative pay-offs). Minus ten represents the negative pay-off of ten years in prison; minus twenty represents twenty years in prison; minus one stands for one year in prison. Zero represents the pay-off of avoiding all the negative pay-offs – that is, of getting off Scot free. I and not-I are the two decisions available to Bill and to Sue: I stands for making the incriminating confession, and not-I for keeping quiet. The pay-off matrix looks like this:

Bill \ Sue	I	not-I
I	$-10 \diagdown -10$	$0 \diagdown -20$
not-I	$-20 \diagdown 0$	$-1 \diagdown -1$

If Bill talks and Sue talks each receives ten years.

If Bill talks and Sue keeps quiet, Bill gets off and Sue receives twenty years.

If Bill keeps quiet and Sue talks, Sue gets off and Bill serves twenty years.

If neither talks, each gets one year.

Here is the dilemma: *the course of solidarity* is to keep quiet. The *safe course* is to talk. However, if Sue takes the course of solidarity and Bill takes the safe course, then Sue has had it; she receives twenty years and Bill is let go.

So perhaps *both* prisoners should do the safe thing, that is, talk. But this would result in ten years for each, much stiffer sentences than they would get had they both been trusting enough to keep silent.

It comes to this, then. *Should Bill and Sue trust to one another's discretion?*

Game Theory offers an answer to this question: a participant should choose so that, no matter what the others do, he is no worse off than before. So Sue should select the decision that, no matter what decision Bill makes, will leave Sue no worse off than before. Thus, if Bill chooses I, and then

(i) if Sue chooses I, then Bill gets ten years (instead of twenty, had he opted for not-I);

(ii)   if Sue chooses not-I, then Bill gets off (instead of getting one year, had he opted for not-I).

So we have it that, for Bill, "the safe course" is best; talking is the strategy that *dominates over* keeping quiet. Talking puts Bill in the best defensive position *no matter what Sue does*.

The trouble is, of course, that exactly the same thing can be said for Sue. *Her* best defensive strategy, no matter what Bill does, is also to choose I. And if Bill and Sue both follow a theoretically dominant strategy – that is, if both talk – then the best *joint result*, in which each gets *just* one year, is out of the question. The individually dominant strategies fail to attain the mutually preferred outcome.

Now what does this show? It shows "that if both players make the rational choice, . . . both lose."[21] Or, to put it another way, "two so-called *irrational* players will fare much better than two so-called *rational* ones."[22]

Is it true that getting the mutually preferred result requires a breakdown of rationality? If so, then why should rationality be a guiding concept of logic, or economics, or of life, if it can pay off so badly? Notice that underlying the Prisoners' Dilemma, *a person's rationality is equated with his own immediate self-interest*. If this equation is correct then, as this paradox shows, sometimes mutual preference requires individual irrationality. But our paradox also suggests another possibility: namely, that the equation of rationality and selfishness is *not* correct.

Let us be clear. The Prisoners' Dilemma *does* show that if each player follows his own selfish interests, then mutual best advantage will be lost. It does *not* show that the concept of self-interest should be identified with the concept of rationality. In fact, the Dilemma suggests the opposite – that the rational thing is not always the most selfish thing. Consequently, any theory, whether of logic or economics, that identifies rationality and self-interest, must be seriously examined; such identification is fallacious.

Notes

1   Quoted in Robert L. Heilbroner, *The Worldly Philosophers* (New York: Simon & Schuster, Revised Edition, 1972), p. 76.
2   Quoted in Leonard Silk, *Economics in Plain English* (New York: Simon & Schuster, 1978), p. 31.
3   Ibid.
4   Ibid., pp. 83–84.
5   Ibid.
6   Physics is another of those sciences. The laws of *microphysics* do not always apply to or have counterparts in the domain of *macrophysics*, and vice versa. In fact, one debate that currently rages in physics is whether the Law of Causality, which is believed to hold for *macro-events*, also holds for *micro-events*.

7   R. Whatley, *Elements of Logic* (New York: Sheldon Press, 1826), pp. 214–215.

8   For a technical answer, see John Woods and Douglas Walton, "Composition and Division," *Studia Logica* 36 (1977): pages 381–406.

9   Ibid., p. 390.

10   Leonard Silk, *Contemporary Economics, Principles and Issues* (McGraw-Hill, Revised Edition, 1975), pp. 13–14.

11   Adapted from Patrick Suppes's article, "The Philosophical Relevance of Decision Theory," *Journal of Philosophy* 58 (1961) pp. 605–614. Much of this section has benefited from conversations with Suppes and from several chapters in Suppes's book, *Studies in the Methodology and Foundations of Science* (New York: Humanities Press, 1969).

12   Ian Hacking, *The Logic of Statistical Inference* (London: Cambridge University Press, 1965), page 186.

13   Jeremy Bentham, *Introduction to the Principles of Morals and Legislation* (London: Athlone Press, 1970), page 26.

14   Actually, the sure-thing principle requires Bill to have a weak preference. The concept of weak preference introduces technicalities that we needn't go into here.

15   John von Neumann and Oskar Morgenstern, *Theory of Games and Economic Behaviour* (Princeton: Princeton University Press, 1944), page 154.

16   R.D. Luce and H. Raiffa, *Games and Decisions* (New York: John Wiley, 1957), page 16.

17   Kenneth J. Arrow, *Social Choice and Individual Values* (New York: John Wiley, 1951).

18   This paradox was examined by E.J. Nanson in a paper in *Transactions and Proceedings of the Royal Society of Victoria* 19 (1882).

19   Vilfredo Pareto, *Manuel d'économie politique*, translated by Alfred Bonnet (Paris: V. Girad & E. Briere, 1909), pp. 617–618 .

20   Nicholas Kaldor, "Welfare Propositions in Economics and Interpersonal Comparisons of Utility," *The Economics Journal* vol. (1939): pp. 549–552; see p. 549.

21   Antol Rapaport, "Escape from Paradox," *Scientific American* 217 (1967): pp. 50–56; see p. 51.

22   R.D. Luce and H. Raiffa, *Games and Decisions* (New York: John Wiley, 1957), p. 91.

Chapter Eleven
Extending Deductive Logic

In the last several chapters, we have seen that the systematicity of formal methods can be applied to the fallacies, provided we are prepared to develop such systems in the appropriate directions. Therefore, it is worthwhile to take a quick look at some of the directions in which formal logic has been developed.

To prepare the way, and because the subject is of interest in its own right, we will first outline the traditional logic of the syllogism, which was, for many centuries, the main type of logic studied. Then we will see how modern formal deductive logic can be extended to deal with the arguments of syllogisms, but with greater generality and depth of treatment. This modern theory is called quantification theory. It is the logic of "all" and "some". Finally, we will look to some other directions in which deductive logic may be extended.

1. Categorical Propositions

We have studied some of the basic concepts of propositional logic. We were concerned with some logical relations between propositions or statements, but we did not analyze the logical structure of the simple propositions (p, q, and so on) themselves. Propositions were taken as the basic, unanalyzed unit of our analysis, and our concern was to elucidate some logical relations that held between propositions. In this chapter, our analysis will cut into the propositions themselves, and we will see the logical structure of some propositions, which we would have regarded as simple propositions up until now.

We will be concerned with *categorical* propositions or statements. Here are examples of the four types of categorical propositions:

A: All politicians are liars.

E: No politicians are liars.

I: Some politicians are liars.

O: Some politicians are not liars.

A and I are *affirmative* propositions, E and O are *negative*. A and E are *universal*, I and O are *particular*.

The forms of the four types of categorical propositions are:

A:	All F are G:	Universal Affirmative
I:	Some F are G:	Particular Affirmative
E:	No F are G:	Universal Negative
O:	Some F are not G:	Particular Negative.

Each categorical proposition is made up of two terms, which are joined by some form of the copula verb "to be". One term is always called the *subject term*, the other the *predicate term*. The subject term is the term that is being affirmed (or denied) as having a certain property. The predicate term is the other term, the one that names the property that is being related to the subject term. In the forms we listed, "F" stands for the subject term and "G" stands for the predicate term.

We will say that a term stands for (or denotes or is true of) a *class* of things. For example, the term "politicians" stands for (or denotes or is true of) the *class* of politicians. We could say that each term has a certain class as its *extension*.

Terms are sometimes nouns, but often adjectives or verbs can also be terms of a categorical proposition. For example, "Some F are G" can stand for both "Some politicians lie" and "Some politicians are people who lie." In another context, "Some F are G" might stand indifferently for "Some fishes fly", "Some fishes are fliers", or "Some fishes are flying things". Complex phrases may also be terms, such as "employed for twenty-seven years by B.F. Goodrich" or "philosophers over thirty who wear Harvey Woods undershirts".

To sum up, the general schema of a standard-form categorical proposition or statement consists of four parts. First the *quantifier*, (that is, "all" or "some" or "no"); then the *subject term*, then the *copula*, then the *predicate term*. The copula is an affirmative or negative form of the verb "to be"; by convention, the present-tense singular or plural form is used.

What we have said so far conforms fairly closely to the traditional, or Aristotelian, interpretation of categorical propositions. But we will now make some assumptions that depart from that interpretation. The account that follows conforms to the modern, or Boolean, interpretation (George Boole, 1815-1864) of standard-form categorical propositions.

First, we will assume that our first proposition, "All F are G", is to be interpreted as meaning "If anything is an F it is also a G." So the statement "All diamonds are gems" will hereafter be read as "If anything is a diamond, it is also a gem." Thus when we assert our first proposition, we are not necessarily asserting the existence of some members of the subject class. So our interpretation, "All diamonds are gems", will not be understood to imply the existence of any diamonds (even though we all know that diamonds do exist).

By saying "All F are G," we do not imply that there are some F. The form says only that "If anything is an F, it is also a G." An example may make this clearer. The statement "All deserters will be shot" does not imply that there are, were, or will be, any actual deserters. It just means that *if* anyone is a deserter, *then* he will be shot. This universal statement is actually a disguised conditional statement. The statement "If anyone is a deserter he will be shot" could be called a universal conditional statement, and it does not imply the existence of any members of the subject class. In general, we will assume that all type A propositions can be interpreted as universal conditionals. So, for our purpose, "All F are G" is to be read "If anything is an F, it is a G." In other words, we will say that the type A proposition does *not* have existential import.

There are many variant ways of stating a type A proposition in English. The proposition "All horses are mammals" could be expressed by any of these sentences: Each (every, any) horse is a mammal. Horses are mammals. If something (anything) is a horse, it is a mammal. If anything is not a mammal, then it is not a horse. All non-mammals are non-horses. Only mammals are horses. No horse is not a mammal. Horses are exclusively mammals.

The second type of proposition has many equivalents. First, we should observe that a type I proposition can always be translated into a type A proposition. For example, "No spiders are insects" can be rendered as "All spiders are non-insects." Other equivalent English forms of the same categorical propositions are: "Nothing is both a spider and an insect"; "Nothing that is a spider is an insect"; "There are no spiders that are insects."

In interpreting type E and type O statements, we will take "some" to mean "at least one". Hence, "Some diamonds are gems" will be interpreted as "At least one diamond is a gem."

In general, common sense and reflection help to best determine the correct standard form of problematic examples. Check-lists of idioms can be misleading, and in some cases tricky. For example, "always" usually means "at all moments" – but not always. The sentence "Walton always dials the phone with his left hand" does *not* mean "Walton dials the phone with his left hand at all moments." The proper interpretation of this example is "All moments at which Walton dials the phone are moments at which Walton dials the phone with his left hand."

We can make the structure of the four types of propositions clearer by using a diagrammatic method devised by John Venn (1880), where circles represent classes. Where a class is empty, it will be shaded. Where a class has at least one member, an "x" will be placed within the circumference of the circle.

In general, four regions can be represented by a diagram of two overlapping circles:

Where F is "French" and G is "generals", the *lens* represents the class of French generals. The *left lune* is the class of French non-generals. The *right lune* is the class of non-French generals. And the area outside both circles is the class of non-French non-generals.

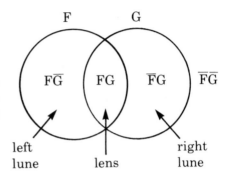

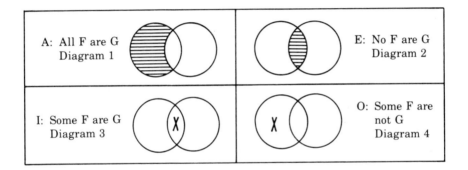

In a Venn diagram, whiteness means nothing but lack of information. In Diagram 2, for example, the lunes are empty, or blank, not because we think that there are some F that are not G or some G that are not F. The lunes are blank because the statement "No F are G" gives us no information about whether there are some F that are not G or some G that are not F.

The symmetry of Diagrams 2 and 3 shows that the terms in E and I can be reversed. That is, "No F are G" is equivalent to "No G are F" and "Some F are G" is equivalent to "Some G are F." But the terms in A and O cannot be switched around. Hence "All Greeks are men" is not to be confused with "All men are Greeks." And "Some men are not Greeks" is not the same as "Some Greeks are not men." This switching of terms was traditionally known as *simple conversion*.

We can also see, from the diagrams, that A and O are mutual *contradictories* or negations. A is true if and only if O is false. Where A has shading, O has an "x", and vice versa. With respect to blankness, or lack of information, they are alike. Clearly A and O simply deny each other. Similarly, E and I are mutually contradictory.

Exercise

1. Identify each of the following statements as A, I, E, or O, and draw the corresponding Venn diagrams.
   (a) There is more than one reporter who is a crime-fighter.
   (b) Everyone who works at the *Daily Planet* lives in Metropolis.
   (c) Whales are mammals.
   (d) No man who changes his clothes in a telephone booth is normal.
   (e) Most dogs are harmless.
   (f) Only dogs who are irrational bite.
   (g) Not every reporter is a disguised crime-fighter.
   (h) Only people over eighteen allowed.

2. Syllogisms

A categorical syllogism is an argument composed of three categorical propositions, which are related in such a way that exactly three terms occur in the argument, and each term appears in exactly two of the propositions.

<blockquote>
All men are mortal.
All Greeks are men.

All Greeks are mortal.

The predicate term of the conclusion is called the *major* term of the syllogism. The subject term of the conclusion is called the *minor* term. The third term, which occurs in both premises but not in the conclusion, is called the *middle* term. Thus "mortal" is the major term, "Greeks" is the minor term, and "men" is the middle term.
</blockquote>

It is a convention of traditional logic that the major premiss (the one that contains the major term) is stated first. A valid syllogism is one whose form is such that it is impossible for the premises to be true and the conclusion false. The Venn diagrams give us a straightforward test for validity.

We will use three intersecting circles to represent the three terms of a syllogism. For a valid syllogism, the conclusion can be seen to be already diagrammed when the two premises are diagrammed.

Consider the example below, letting G = men, H = mortal, and F = Greeks.

All men are mortal	All G are H
All Greeks are men	All F are G

∴All Greeks are mortal.   ∴All F are H.

We represent the first premiss, a type A proposition, by shading the area in G that lies outside the circumference of H.

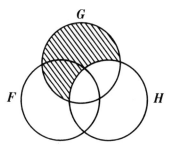

Then we represent the second premiss, also an A proposition, by shading the area in F that lies outside G.

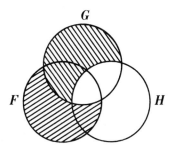

We can see that the conclusion, All F are H, is already represented on the diagram: all the area in F that lies outside the H area is shaded. Hence we conclude that the argument is valid.

Let's try another example. Let F = Greeks, G = philosophers, and H = wise men.

All philosophers are wise.	All G are H
Some Greeks are philosophers.	Some F are G

Some Greeks are wise.   Some F are H

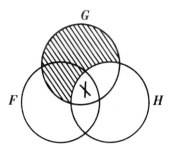

This example also illustrates why the universal (A or E) premiss should always be inscribed on the diagram before the particular (I or O) premiss. If we had inscribed the second premiss – Some F are G – on the diagram first, we would have had to resort to the device of inscribing it on the line that divides the lens area between F and G. This is necessary because the "x" only signifies that the class represented by the F-G lens area has at least one member. It does not specify whether this member lies in the upper leftmost lens area, or in the lower rightmost lens area. All we know is that the class represented by the lens area has at least one member that may fall within either the upper or lower area (or both). To represent this, we would inscribe the "x" on the line:

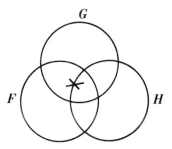

However, if we inscribe the universal premiss – "All G are H" – first, we see that the upper lens area between F and G is empty.

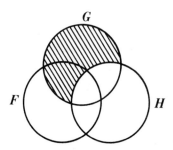

So our need to resort to the strategy of putting the "x" on the line is obviated. We now know that the "x" can only go in the lower area of the

F-G lens. The general rule is therefore: always inscribe the universal (A or E) premisses first.

Here is an example where the Venn diagram demonstrates invalidity. Let F = Muslims, G = Mennonites, and H = pacifists.

| All Mennonites are pacifists. | All G are H |
No Muslims are Mennonites.	No F are G
No Muslims are pacifists.	No F are H

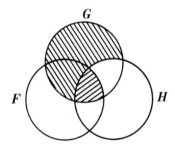

This argument is judged invalid since the conclusion, "No F are H", is not automatically inscribed on the diagram by the inscription of the premisses. Observe that the lens area between F and H is not entirely shaded, so the proposition "No F are H" is not represented.

One more example will illustrate the strategy of putting the "x" on the line. Let F = Freudians, G = Gestalt psychologists, H = hot-tempered.

| Some Gestalt psychologists are not hot-tempered. | Some G are not H |
Some Freudians are not Gestalt psychologists.	Some F are not G
Some Freudians are not hot-tempered.	Some F are not H

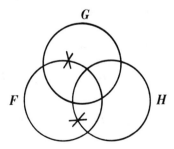

Notice that in inscribing "Some G are not H", we put the "x" on the line within the G-circle, because the premiss tells us only that there is at least one thing within the G-class and outside the H-class. The premiss tells us nothing about the F-class. To indicate this lack of knowledge we follow one convention of putting the "x" on the line, which is part of the circumference of the F-circle. The same strategy is used in inscribing the second premiss. Now we can see that the conclusion is not inscribed on the diagram, as we do not know whether the class represented by the F-H lens has any members since the lower "x" is merely on the line and not definitely within the F-H lens area. Hence "Some F are not H" is not inscribed, and the syllogism is invalid.

A so-called "Square of opposition" can be drawn to illustrate the fact that A and O, as well as I and E, are contradictories.

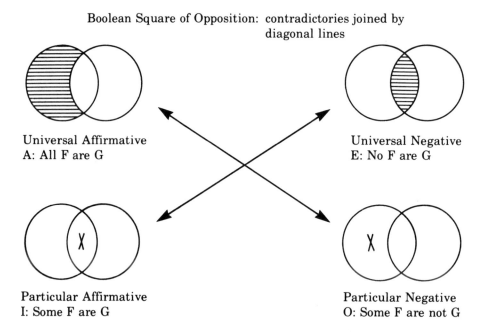

Boolean Square of Opposition: contradictories joined by diagonal lines

Universal Affirmative
A: All F are G

Universal Negative
E: No F are G

Particular Affirmative
I: Some F are G

Particular Negative
O: Some F are not G

Two propositions are *contradictories* if it is not logically possible for them both to be true, and if, in addition, it is not logically possible for them both to be false. Two propositions are *contraries* if it is not logically possible for them both to be true, even though it may be possible for them both to be false. For example, "Leo is yellow" is the contradictory of "Leo is not yellow", but only the contrary of "Leo is purple".

3. How to Set up a Venn Diagram Test for Validity

1. Set out the form of the argument. There must be two premises and one conclusion, and all three of these propositions must be cast into

standard form, that is, type A, I, E, or O. There must be exactly three terms in the argument, and each term must appear in two of the propositions. When presenting the argument form, symbolize each term by a capital letter.

2. Construct three overlapping circles to represent the three terms of the argument, and label each circle with a letter that stands for a term.

3. Inscribe the contents of the two premisses onto the diagram. *Always* inscribe the universal premiss first, if one premiss is universal.

4. Notice that sometimes an "x" must be placed on the line instead of inside an area. This must be done when we want to indicate that we don't know which of two areas something is in. In our diagram, the left side states that there is something in (the class represented by) the left circle where we don't know whether it is in the left lune area or in the lens area or both. Next, consider the example on the right.

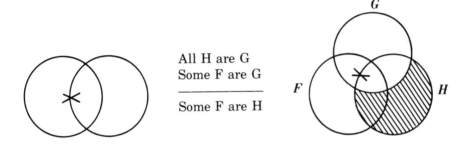

All H are G
Some F are G
_____
Some F are H

Inscribe the universal premiss first by shading the lune of H. Then the second premiss must be inscribed by placing the "x" *on the line* within the lens between F and G. This indicates that we do not know whether there is something specifically in the upper (leftmost) area of the lens or the lower (rightmost) area. The second premiss tells us only that something is *in the lens* between F and G; we do not know, according to the second premiss, whether it is in the upper part of the lens, or the lower part, or both. All we know is that it is in the upper part, or in the lower part, or in both. Clearly, the conclusion, therefore, is not inscribed on the diagram, since we do not actually know that there is something in the upper (unshaded) area of the lens between F and H. Obviously, it is possible that this lens area is empty, since the "x" is only on the line and not completely within the unshaded lens area between F and H.

5. See whether the content of the conclusion is automatically inscribed on the diagram as a result of step 3. If so, the argument is valid; if not, the argument is invalid. In our example, the argument is invalid.

Exercise

1. Symbolize the following arguments. In each case, show what

the symbols represent, and show whether each argument is valid with the aid of Venn diagrams.

(a) All Communists are socialists, so all Hungarians are Communists, as all Hungarians are socialists.

(b) There are a few actors who are not egomaniacs, but the genuine thespian is, without exception, an egomaniac. Hence all genuine thespians are actors.

(c) All Reds are Commies but no fellow travelers are Reds, so no fellow travelers are Commies.

(d) There's no such thing as a delinquent who isn't maladjusted. Also, many rich kids are delinquents. So there are some rich kids who are maladjusted.

(e) There's no such thing as an ambitious mathematician, since only ambitious people are politicians and no mathematicians are politicians.

(f) Some parrots aren't pests and all parrots are pets so no pets are pests.

(g) Some glargs are not harbs and some fangs are not glargs, so some fangs are not harbs.

(h) Not all people who are irrational are illogical, since nobody who is illogical is confused and many people are both irrational and confused.

(i) No artists are rational, since no painters are rational and no artists are non-painters.

THIS QUESTION IS OPTIONAL:

2. Symbolize and determine the validity or invalidity of the arguments using Venn diagrams.

(a) If a man has sweaty palms, he is a liar. There never was a liar who didn't tremble when confronted by a crocodile. Therefore, if a man has sweaty palms, he is certain to tremble when confronted by a crocodile.

(b) Babies are illogical. Nobody who can manage a crocodile is despised. Illogical persons are despised. Therefore babies cannot manage crocodiles.

(c) All men are basically cowards; hence it follows that Jones is a coward.

4. Quantifiers

Quantification theory is the modern theory of concepts corresponding to the English words "all" and "some". The symbol ($\forall$ x) is read as "for all x . . .". The sentence ($\forall$ x) Fx, in quantification language, is interpreted as "For all x, x is an F" or, in other words, "All x's are F's", or "Everything is an F'". The existential quantifier may be defined in terms of the universal quantifier and negation.

$$(\exists x)\ Fx \underset{df}{=}\ \daleth (\forall x)\ \daleth Fx$$

$(\exists x)$ Fx is read as "Some x are F'" or (more exactly) "At least one x is an F'". $(\exists x)$ Fx says the same thing as "It is not the case that all x's are not F's", or "Not all x's are not-F's". In other words, this definition expresses an equivalence between "Something is an F'" and "Not everything is not-F'".

Existential quantification may be thought of as infinite disjunction; universal quantification may be thought of as infinite conjunction. An existential statement, $(\exists x)$ Fx, means that there is at least one thing in the universe that has property F. Suppose the universe consists only of a finite set of n objects, $a_1, a_2, \ldots a_n$. Then $(\exists x)$ Fx, or "At least one object has F'", means exactly the same thing as: $Fa_1 \lor Fa_2 \ldots \lor Fa_n$. Either object a has F *or* object a_2 has F *or* object a_3 has F, and so on, to a_n. Similarly, to say $(\forall x)$ Fx (Everything in the universe has F) is simply to say that a_1 has F *and* a_2 has F *and* a_3 has F, and so on to a_n. In fact, universal quantification could be defined in terms of conjunction, and existential quantification in terms of disjunction, thus making a separate theory of quantification unnecessary. But this would be undesirable, as there are reasons for not allowing infinite series in truth-functional logic, whereas we often want to quantify over infinite universes, such as the class of integers.

Translation of the four categorical propositions into the notation of Quantification Theory is done as shown on the Square of Opposition below.

QUANTIFICATIONAL SQUARE OF OPPOSITION

A: All F are G.
$(\forall x)\ (Fx \supset Gx)$

I: Some F are G.
$(\exists x)\ (Fx \land Gx)$

E: No F are G.
$(\forall x)\ (Fx \supset \daleth Gx)$

O: Some F are not G.
$(\exists x)\ (Fx \land \daleth Gx)$

Notice that the quantificational interpretation of the A-proposition is clearly a universal hypothetical and does not assert the existence of members of the F-class. $(\forall x)\ (Fx \supset Gx)$ reads, "If anything is an F then it is a G." We can use the interdefinability of the quantifiers plus some truth-functional equivalences to make quite explicit the fact that the diagonal lines indicate contradictoriness (negation) as follows. We know that the A-proposition, $(\forall x)$ $(Fx \supset Gx)$, is equivalent to $\daleth(\exists x)$ $\daleth(Fx \supset Gx)$ because $(\forall x)$ Fx is always equivalent to $\daleth(\exists x)$ $\daleth Fx$ no matter what expression we put in for Fx. But since we can show, by a truth-table, that $\daleth(p \supset q)$ is equivalent to $(p \wedge \daleth q)$, we can substitute $(Fx \wedge \daleth Gx)$ for $\daleth(Fx \supset Gx)$ in $\daleth(\exists x)$ $\daleth(Fx \supset Gx)$, getting $\daleth(\exists x)$ $(Fx \wedge \daleth Gx)$. So we know by this reasoning that $(\forall x)$ $(Fx \supset Gx)$ is actually equivalent to $\daleth(\exists x)$ $(Fx \wedge \daleth Gx)$. Also, intuitively, this equivalence holds as "All marbles are round" is equivalent to "There are no marbles that are not round." But observe that our new transform of the A-proposition, $\daleth(\exists x)$ $(Fx \wedge \daleth Gx)$ is simply the negation of the O-proposition, $(\exists x)$ $(Fx \wedge \daleth Gx)$. Hence the A and O-propositions are negations or contradictories. By similar equivalences it can be shown that E- and I-propositions are contradictories.

Exercise

Show, by a series of equivalences similar to the above deduction, that the type E and type I propositions are contradictories in quantification theory.

Having gotten some idea of how quantifiers enable us to express the basic ideas behind syllogistic arguments, let us now see how quantification theory could be developed as an extension of system S.

Here is one way of looking at our system S: some deductions are good, some inferences are warranted, and some entailments obtain mainly because of the meanings of the expressions $\daleth$, $\vee$, $\wedge$, $\supset$, $\equiv$, and independently of the meanings of the sentences or formulas of which they are connectives. Consequently, we may speak of $\daleth$, $\vee$, $\wedge$, $\supset$ and $\equiv$ as *trigger-terms:* they are terms that trigger the success or failure of deductions and inferences in S. Now it is a fair (and good) question to ask whether there are other trigger-terms in the logic of deduction. The short answer is that there are.

Not every good deduction is triggered by words like "not-", "or", "and", "if . . . then" and "if and only if" (or their counterparts in S). Quite commonly, a deduction will be triggered with the help of such expressions as "all" and "some", or by other terms in some other formal systems. Let us briefly sketch such a system.

5. The System Q of Deductive Quantificational Logic

The system Q uses S as its base. That is, it absorbs S and is an *extension* of it. In addition to sentence variables and sentence constants, the system Q adopts four further categories of expression:

(1) *Quantifier constants*
Ɐ, which is pronounced "all" and
Ǝ, which is pronounced "some".

(2) *Predicate variables*
F, G, H, F_1, F_2, F_3, . . .

(3) *Singular subject variables*[1]
Name variables[1]: a, b, c, a_1, a_2, a_3, . . .
Pronoun variables: x, y, z, x_1, x_2, x_3, . . .

(4) *The special name constant, δ.*

A word or two of explanation is in order. It may be asked why we use the two special symbols, Ɐ and Ǝ, rather than their plain English counterparts. It is useful to remember that, in S, there was not an exact equivalence of meaning between the connectives and their counterpart English expressions. Moreover, we opted to use the connectives in S, rather than the English expressions, because the connectives have a more interesting logical structure, because they have a systematic truth-functional structure. In effect, in system S we made a slight regimentation, which was justified by a gain in logical clarity. The same holds true of Ɐ and Ǝ. Let us see why.

In English, it might seem that the form "All F are G" *certainly* entails the form "Some F are G". And in English, it might seem that "Some F are G" entails "Two or more F are G". From a strictly logical point of view, however, we want "some" to have an *exact and unvarying* meaning. So it makes sense to interpret "some" in the sense of "there exists at least one". No logician wants to insist that in English this is what "some" invariably means. Logic has no imperial designs upon English. Instead, it is proposed that a new symbol be introduced; that symbol is *given* the meaning "there exists at least one". Even if the new symbol does not have precisely the same meaning as "some" in English, the meaning of the new symbol will certainly very closely resemble the English meaning. As before, we are speaking here of *explication* of common-sense English.

But if Ǝ is going to mean "there exists at least one", can we strictly speaking allow it to imply "there exists at least one"? For are not some laws of science perfectly accurate laws, even though they discuss a thing that does not exist? For example, look at the law, "All dodos have a functionless wing-structure." Is the law inaccurate because there is no such thing as a dodo? And if there aren't any dodos, how can it be true that there exists at least one dodo? So it would seem that Ɐ, in this sense, cannot imply "some" in the same sense that Ǝ implies "some". Consequently, in Q, we introduce the expression Ɐ, and we give it a refined meaning that permits us to make statements about dodos without embarrassment. Once again, although Ɐ does not mean "all" in common-sense English, it comes very close to it.

By the way, the quantifier constants of Q (or "quantifiers", for short) are so-called because they roughly indicate the numbers of things being talked about.

In English, we have good inferences that do not depend upon truth-functional structure. For example:

> All human beings are mortal.
> Socrates is a human being.
> _____
> ∴ Socrates is mortal.

If we were to represent the logical form of this deduction in S, we would have something like this:

> *p*
> *q*
> ———
> ∴ *r*

The form is certainly not deductively correct. But it is easy to see that, though our deduction is correct, and correct by virtue of its logical form, the form, namely,

> All F are G
> a is F
> _____
> ∴ a is G

we are speaking, nevertheless, about a form of a different sort. Simple inspection of

> All F are G
> a is F
> ∴ a is G

shows that the expressions that don't matter in the deduction

> All human beings are mortal.
> Socrates is a human being.
> ∴ Socrates is mortal.

are not constituent *sentences* but rather constituent sentence *particles*, namely the *predicate* expressions "is a human being" and "is mortal" and the *subject* expression "Socrates". Consequently, we shall follow our usual practice of using *variables* in place of those types of expression the meanings of which do not count for the correctness of the deduction in which they occur. And, so, we have *predicate variables* and *subject variables*.

There are two types of subject variables. In English, we sometimes reason:

> If anyone comes, he will be welcome.

Now here there is no specific reference to any *named* person. But not all reasoning is unnamed; for example:

> If Charlie comes, he will be welcome.

In English, unspecific, unnamed reference is achieved by various kinds of pronominal expressions, of which the pronoun is possibly the best known and most widely used. And so it is in Q, where we would represent the first sentence by (loosely speaking) something like:

> For all x, if x comes then x will be welcome.

The second sentence (again, loosely speaking), would be represented by something like:

> If C comes, then C will be welcome.

The special name constant, δ, has a function that is quite easily explained: δ is allowed to be a name, admittedly a special kind of name; δ names *everything whatever*.

Some deductions are correct by virtue of their logical forms. For example:

> All F are G
> a is F
> _____
>
> ∴ a is G

Such deductions are said to be correct by virtue of their *quantificational deductive form* (or, for short, quantificational form). The system Q of quantificational theory studies the logically interesting properties of *quantificational forms*.

There are two more important points about the logic of quantificational form. First, we shall show how the system Q solves the problem of making true general assertions about the non-existent, assertions such as "All dodos have a functionless wing-structure." And second, we shall list examples of some of the more important deductively correct logical forms of Q.

Making true general assertions about the non-existent is known technically as *quantifying over empty domains*. In Q, our dodo-assertion has the following logical form:

$$(\forall x)\,(Fx \supset Gx)$$

In other words, "For all things x, if x is a dodo, then x has a functionless wing-structure." Notice that we do not assert that anything is a dodo; we say only that *if* anything is a dodo, then it also has a functionless wing-structure.

SAMPLES OF DEDUCTIVELY CORRECT LOGICAL FORMS OF Q

Since this formula is always true in Q,	this deduction is correct,	and this inference is warranted.
$(\forall x)(Fx \supset Fx)$	$(\forall x)(Fx)$ $\therefore Fx$	from $(\forall x)(Fx)$ it is warranted to infer Fx.

The law of Q by which this is so is called the Law of *Universal Instantiation*.

$(F\delta) \supset (\forall x)(Fx)$	$F\,\delta$ $\therefore (\forall x)(Fx)$	From $F\,\delta$ it is warranted to infer $(\forall x)(Fx)$

The law of Q by which this is so is called the Law of *Universal Generalization*.

$(\exists x)(Fx) \supset Fa$	$(\exists x)(Fx)$ $\therefore Fa$	From $(\exists x)(Fx)$ it is warranted to infer Fa.

The law of Q by which this is so is called the Law of *Existential Instantiation*.

$(Fa) \supset (\exists x)(Fx)$	Fa $\therefore (\exists x)(Fx)$	From (Fa) it is warranted to infer $(\exists x)(Fx)$.

The law of Q by which this is so is called the Law of *Existential Generalization*.

To see how the system Q provides a precise means of dealing with some fallacies of equivocation, we will look back to an earlier example of equivocation.

> (E) Everything has a cause.
> Therefore, there is some thing that causes everything.

Used thus equivocally, as we saw, the premiss might provide a convincing argument for the existence of One Cause, or God. But the quantificational structure of (E) displays two arguments. Let a be a specific individual, and let Cab be a predicate for "a causes b".

(E1)   (∀x) (∃y) Cyx
_____

(∃y) (∀x) Cyx

(E2)   (∀x) Cax
_____

(∃y) (∀x) Cyx

The classical pattern of equivocation is evident. (E1) is invalid, but the premiss is plausible. (E2) is valid, but the premiss is implausible. The equivocator seeks both the true premiss and valid argument; following the classical pattern of equivocation, he effects a binocular blurring of (E1) and (E2) into the appearance of one argument, (E). Q clearly displays the error.

6. Some Further Developments of Deductive Logic

The trigger-words of S are ℸ, ∨, ∧, ⊃, ≡. The trigger-words of Q are ∀, ∃ and δ. Might we not expect to find still further trigger-words in the logic of deduction? Yes, indeed. For example, in making inferences in English, we quite commonly assert connections between what is *necessary* and what is *possible*. For example: if we think that humans are necessarily rational (as benign Aristotle believed), then we certainly would not want to deny that they are *possibly* rational. Similarly, sometimes we assert deductive connections between what is *necessary* and what is *necessary*, as in this example:

> All humans are necessarily rational.
> ∴ No humans are necessarily not-rational.

The expressions "it is necessarily the case that"("necessarily") and "it is possibly the case that" ("possibly") are called *alethic modal operators*. They are expressions which, when applied to sentences, form new sentences (hence the term "operator"). For example, suppose the operator "possibly" is applied to this sentence:

> Bill is a fool.

There is formed a second sentence:

> Possibly Bill is a fool.

Furthermore, "necessarily" and "possibly" are expressions that indicate the *mode* of truth or falsehood of the propositions to which they apply (hence the term "modal"). Hence, too, the term "alethic", which means "having to do with truth and falsehood."

Just as some deductions are correct by virtue of truth-functional logical form, and others by virtue of quantificational logical form, still others will be correct by virtue of their *modal logical form* (or, more briefly, their modal form). For example, the deduction about Bill, just a few lines back, exhibits this perfectly general (and correct) modal form:

$$p$$
$$\therefore \text{ possibly } p$$

The theory of deduction contains, therefore, a modal component, an *extension* of S and Q. Here, very briefly, is a sketch of just one of several modal extensions of S; it is called "the system M".

The System M In addition to the basic vocabulary of S, M makes use of two *modal constants:* L, pronounced "necessarily", and M, pronounced "possibly". The letter L is a reasonable choice as the symbol for "necessarily", inasmuch as it suggests "logical necessity"; and the choice of M for "possibility" is derived from *möglich*, the German word for "possible".

Here are some of the characteristic deductive patterns of the system M:

> *Example 1:*
>
> Lp
> $\therefore$ p
>
> *Example 2:*
>
> Mp
> $\therefore$ ⌐ L ⌐ (p)
>
> *Example 3:*
>
> L ($p \supset q$)
> $\therefore$ L$p \supset$ Lq
>
> *Example 4:*
>
> p
> $\therefore$ Lp, where p
> is a theorem.
>
> *Example 5:*
>
> p
> $\therefore$ Mp
>
> *Example 6:*
>
> ⌐ M ($p \lor q$)
> $\therefore$ ⌐ M$p \land$ ⌐ Mq

Example 7:

M $(p \wedge q)$
∴ Mp $\wedge$ Mq

Example 1 shows that the necessity of a proposition implies its truth. Example 2 tells us that the possibility of a proposition is incompatible with the necessity of its negation. Example 3 tells us that the modal operator distributes through the connective $\supset$. Example 4 tells us that any theorem of M is also a necessary truth of M. Example 5 tells us that any true proposition is also possible. From example 6, we learn that if a disjunctive proposition is not possible, then neither of its disjuncts is possible. And example 7 tells us that the operator, M, distributes itself through the connectives for conjunction. Note that the converse of 7 does not hold. The propositions "Socrates is black" and "Socrates is white" are each individually possible. But their conjunction is not possible.

As we have said, M is not the only modal extension of S. Two other well-known extensions are the systems S4 and S5, originally devised by the American logician, C.I. Lewis. The names S4 and S5 indicate that Lewis had devised a series of modal systems, which he identified by consecutive numbers together with the letter S.

It is especially interesting to note that, in S4, the necessity operator, L, is allowed to *iterate*, as logicians say. That is, in S4 the following is a deductively correct modal form:

Lp
∴ LLp

In S5, we get a somewhat more complex sort of iteration. In S5 the modal form

Mp
∴ LMp

is deductively correct. These two logical forms jointly suggest that, in some systems of modal logic, formulas will have whatever modal status they have as a matter of necessity.

Many fallacies occur when an arguer gets the *scope* of a modal operator wrong. L $(p \supset q)$ means that the whole hypothetical "If p then q" is a necessary truth, or that p entails q. But $p \supset$ Lq means that if p is true then q is a necessary truth. For example, "Bananas are yellow and apples are red" entails "Bananas are yellow". So necessarily, if the first proposition is true, so is the second. Here L $(p \supset q)$ is appropriate. But assume that the proposition "Bananas are yellow and apples are red" is true; the proposition does not warrant the conclusion that "Bananas are yellow" is necessarily true. Quite the contrary – because the second statement is contingent. It could be false that bananas are yellow. So $p \supset$ Lq is not appropriate.

The point may seem simple, so explained, but the ambiguity of the common idiom "If p then it must be that q" in English makes it difficult to understand what is meant.

For example: $p \supset p$ is necessarily true in S. Consequently, by the rule of example 4, we may infer that L $(p \supset p)$. An instance: necessarily, if Jones dies on Tuesday, then Jones dies on Tuesday. But we must not infer $p \supset Lp$. An instance: if Jones dies on Tuesday, then necessarily Jones dies on Tuesday. The second statement seems to imply that Jones must die on Tuesday, whatever else might happen – a fatalistic suggestion.

Exercise

1. The inference "L $(p \lor \neg p)$, therefore $Lp \lor \neg p$" is *not* correct in modal logic. Can you find some arguments in English that fallaciously trade on a scope confusion of this sort?
2. Discuss this argument:
   It is possible for me to walk.
   It is possible for me to sit.
   ———————————————————————
   It is possible for me to walk and sit.

A quick look at a system of quasi-deduction A system of "quasi-deduction" is a system in which the connection between premisses and conclusion is *not always strong enough* to justify the claim that premisses and conclusion form a *deductively correct* argument (or a deductively correct argument form). Strictly speaking, systems of quasi-deduction allow arguments that would not be correct from the point of view of deductive logic. Possibly this sounds illogical, or at the very least over-generous, but we think that a few examples will indicate that quasi-deductive logic is not a frivolous enterprise. First of all, we introduce two new kinds of operators, the *epistemic and doxastic* operators. The epistemic operator, K, pronounced "it is known that", is the main operator of "epistemic logic", the logic of knowledge. The word "epistemic" comes from the Greek word *episteme*, which means "knowledge". When applied to a sentence, our operator generates a new sentence. For example, we have the sentence

Bill is a fool.

The application of our epistemic operator, K, would give an additional sentence:

It is known that (Bill is a fool.)

Our doxastic operator is B, and is pronounced "it is believed that". The word "doxastic" comes from the Greek, meaning "of, pertaining to, or depending upon belief." B is also a sentence operator. For example, if we have the sentence p, then the application of the operator p generates the new sentence, Bp.

Among the correct logical forms of epistemic and doxastic logic, we find these two:

>*Example 1:*
>
>Kp
>$\therefore$ p
>
>*Example 2:*
>
>Kp
>$\therefore$ Bp

We learn, in the first example, that knowledge implies truth; that is, you cannot know what is not true. And example 2 tells us that knowledge implies belief; that is, if you do not believe a thing, you can hardly be said to know a thing. It would appear, in fact, that these two argument forms are *deductively correct*. But what of those logical forms that would not be deductively correct? What of forms that are only considered to be correct as quasi-deductions at best? Here are some examples:

>*Example 3:*
>
>Kp
>$\therefore$ KKp
>
>*Example 4:*
>
>Kp
>$p \supset q$
>$\therefore$ Kq
>
>*Example 5:*
>
>Bp
>$p \supset q$
>$\therefore$ Bq
>
>*Example 6:*
>
>Bp
>$\therefore$ BBp

Example 3 tells us that if one knows, one always knows that one knows. Now, this certainly sounds like a reasonable conjecture, but it surely depends upon how aware and reflective we are about our own states of

knowledge. *It also depends on how we are prepared to define knowledge!*
One very ancient definition of knowledge stems from Plato:

> S knows that *p* if and only if:
> 1) *p* is true
> 2) S believes that *p*
> 3) S has good reason to believe that *p*

If this is a correct definition of knowledge, then it is entirely possible
that one could know *without* knowing that one knows. For example,
suppose that I believe some proposition, A, and suppose that I believe
that proposition on the evidence E. It is quite possible that I may doubt
that evidence E completely or sufficiently justifies my believing A. But
what if I am wrong about this? What if, unknown to me, evidence E is
perfectly sound justification for believing A? And what if, in spite of my
own marginal doubts or reservations, E is true? Then it would be the
case that A is true, that I believe that A is true, and that I have adequate
reasons for believing that A is true. Even though I didn't think I *had*
adequate reasons for believing that A, nevertheless (by this definition) I
would know that A is true, yet certainly would not know that I knew.
However, it may be that in some kind of ideally rational world, we would
restrict our knowledge claims to those claims that are not only correct,
but which we also know to be correct. Under such idealized conditions,
therefore, example 3 would be an example of a good deduction. So we
might characterize a quasi-deduction as an argument which, though
deductively incorrect, would be deductively correct under certain
idealized rationality assumptions.

Under suitably idealized conditions of rationality and reflectiveness,
example 6 would, we suppose, be correct – we could say that if we
believed a proposition, then we believe that we believe it. And example 5
and 4 indicate that, in a perfectly rational world, it would be the case
that we believe all the consequences of what we believe, and that we
know all the consequences of what we know. In such a world, if I knew
that *p*, and if *p* implied *q*, then I would believe that *q*. Otherwise, I might
find myself in a situation in which I asserted that *p* (because I believed
it) and denied that *q* (because I didn't believe it). I would have
contradicted myself. In an ideally rational world, there would be no
place for contradiction; knowledge and belief would be allowed to have
the features claimed for them in examples 4 and 5. Technically
speaking, example 4 says that knowledge is *closed under consequence*;
example 5 says that belief is *closed under consequence*. These, then, are
good *quasi-deductions*. Though it is not *in fact* true that a person always
believes or knows all the consequences of what he believes or knows, it
seems safe to suppose that, in an ideally rational world, he would.

Quasi-deduction helps to clarify the structure of the *ad ignorantiam*
fallacy. This fallacy is involved in one or the other of the pair of
arguments:

(1) Nobody has ever proved that *p* is true.
Therefore, *p* is true.
(2) Nobody has ever proved that *p* is false.
Therefore, *p* is true.

In an epistemic logic, ⅂ K*p* ⊃ ⅂*p* is not a tautology, so argument (1) is a fallacy. The fallacious move can possibly be explained as an illicit shift of the negation sign, for K ⅂ *p* ⊃ ⅂ *p is* a tautology of quasi-deduction. It is simply a substitution instance of K*p* ⊃ *p*, which *is* a tautology. As we saw, "K*p*, therefore *p*" is a correct argument in quasi-deduction.

Exercise

Take the second form of *ad ignorantiam* argument (argument 2) and show how epistemic logic could clarify the structure of the fallacy.

7. Summary Remarks about the Fallacies

There are really only two points to be made here. One is that the theory of deduction is perfectly tolerant of some of the most serious abuses of reason. As we have said before, the deduction "p, ∴ p" is a paradigm of sound deductive reasoning, yet it is also a paradigm of bad argument, for it commits the fallacy of arguing in a circle, or *petitio principii*. Even if we mastered all the laws of deduction, and even if we complied with them faithfully in our own argumentative moments, we would *not* be guaranteed to stay free of all the fallacies.

The second point to notice is that though some of the fallacies are bad deductions, the theory of deduction does not *explain* or *account for* the precise respect in which the fallacies are fallacious. Consider, for example, the fallacy of the *circumstantial ad hominem*. When you commit this fallacy, you base your conclusion on certain observations about your opponent. Even if these observations are true, they do not warrant the conclusion in question. Now such an argument will be interpreted, by deductive logic, to be a bad deduction, to be a deductively incorrect argument. But there is much more to the fallacy of the circumstantial *ad hominem* than deductive incorrectness. The theory of deduction lacks the capacity to point out that there is a specific kind of *reason* that a fallacious *ad hominem* argument is a bad argument, or why its conclusion is not entailed by its premises. Without dialectic, deductive logic does not even have the concepts necessary to offer this explanation.

Given that our principal concern is to provide a description and evaluation of the logical structure of the fallacies, it follows that we must advance beyond the province of deductive logic. This is not to demean deductive logic. It is only to say that deductive logic yields a limited perspective. Its virtues are impressive virtues indeed, but there are virtues deductive logic does not have.

Summary

We now want to review our explanations of the following notions:

— logic as a *theory*
— logic as a *formal theory*
— the logic of the fallacies as *informal* logic

(1) Logic as a Theory

Logic is the study of reasoning. It preserves what we already know about reasoning. It exposes underlying principles in a systematic and orderly way. It uses such principles to make new discoveries. And it pursues its objectives of systematicity and orderliness with the help of a refined vocabulary of special terms, which are explications of ordinary, common-sense terms. The explicating terms that we have already encountered here are: ⅂, ∨, ∧, ≡, →, ⟷, ∀, ∃, δ, L, M, K and B.

(2) Logic as a Formal Theory

A logical theory is formal to the degree to which it analyses the structure of reasoning by developing a theory of *logical form*. Moreover, a logical form is formal to the extent to which variables dominate over constants. Some of the logical forms that we have already dealt with are:

$$p$$
$$\therefore p$$

$$p$$
$$p \supset q$$
$$\therefore q$$

$$(\forall x)\,(Fx)$$
$$\therefore Fx$$

$$Fx$$
$$\therefore (\exists x)\,(Fx)$$

$$Mp$$
$$\therefore \; ⅂\, L\, ⅂\, p$$

$$p$$
$$\therefore Mp$$

$$Kp$$
$$\therefore Bp$$

$$Kp$$
$$\therefore KKp$$

(3)   The Logic of the Fallacies as an Informal Theory

It is quite common for logicians to speak of the theory of the fallacies as constituting a branch of *informal logic*. But it is not clear what is meant by the term "informal logic" in this context. Sometimes it seems to be suggested that informal logic encompasses any branch of logic that falls outside the theory of deduction. If this is what we agree to mean by "informal logic", then it is true, as we have several times insisted in this book, that the theory of the fallacies is indeed informal.

However, when one thinks about it a bit, there is no very good reason to reserve the title of formal logic for the theory of deduction. We have already seen that formality is a property that, in varying degrees, may be claimed for branches of logic other than deductive logic. For example, *probability theory*, which we looked at in Chapter Four, is a formal theory. It makes heavy use of variables, and its configurations of variables and constants yield up identifiable logical forms. So we do not accept the interpretation of "formal" that would make formality the exclusive property of deductive logic.

Another interpretation is sometimes suggested for the concept of formality. Some logicians write as if they believe that a theory is formal precisely to the extent to which it is systematic, orderly, and capable of the most general kinds of statement. That is, formality coincides exactly with degree of theoretical development: the more powerful and more settled and more complete the theory, the greater its formality.

Now there is no doubt that theoretical maturity and formality often go hand in hand. But it is also true that some theories – admittedly not very elaborate ones, perhaps – can be developed to a point of very considerable maturity and generality with little or no use of formal methods. In other words, mature theoretical accomplishment is *sometimes* possible in a language in which there are very few variables and in which there is very extensive reliance on the full meanings of most of the sentences of such a language. Accordingly, we do not think it appropriate to recommend this interpretation of "formal theory".

In this book, we have taken the view that theory is formal to the extent that it devotes itself to matters of form. If the theory is a logical theory, then its formality will be a function of the extent to which it deals with logical form. Moreover, the involvement of a theory with logical forms will be more or less a formal involvement, depending upon the extent to which, in those logical forms, variables dominate over constants. This, we will suggest, is a sound and useful interpretation of the notion of formality. It provides that deductive logic is formal logic, and that the quasi-deductive systems of epistemic and doxastic logic are also formal theories. But it *also* allows us to say that deductive sentence logic is *more* formal than epistemic logic, since epistemic logic makes essential use of a greater range of constants.

This is our interpretation of the notion of formality. It can now be asked whether the theory of the fallacies will be a formal theory. The answer is entirely straightforward. *Of course it will be a formal theory.* For example, the theory of the fallacies will certainly proclaim that the argument form "p, ∴ p" commits the fallacy of *petitio principii*. Here there is reference to a logical form, and the logical form in question is one in which the dominance of variables to constants is at its most extreme – for there are no constants! Having made this point, it is also useful to remark that, in our analysis of certain of the fallacies, it is inevitable that we will expose, for consideration, logical forms in which there is considerably higher incidence of constants than in the example we have just mentioned.

> The theory of the fallacies, then, will be a more or less formal theory depending upon the formal complexity of the logical structures that it investigates.

Notes

1 Monadic quantificational logic results from using only single-variable predicate letters. For example, Fx (x is yellow) uses only a single variable, whereas Fxy (x is east of y) uses more than one variable. All of monadic quantificational logic can be formalized using only a single variable. Because we have many-variable predicate letters in mind, we use a multiplicity of variable letters.

Appendix I:
A Sample Dialectical Disputation

This sample disputation will serve as a realistic example of the dialectical model of argument outlined in Chapter Six. The particular subject-matter chosen is not important, provided the issue is clearly enough formulated to allow the model of dialectical argument to obtain a foothold, and provided the topic is one lively enough to encourage an opposition of theses that will generate arguments on two or more sides. In this instance, we chose two theses that are directly opposed to each other.

THESIS A: Cannibalism is always morally wrong in any civilized society.

THESIS B: Cannibalism is not always morally wrong in any civilized society.

We have two participants in our game of dialectic: Arthur will defend THESIS A and Betty will defend THESIS B. In the first round, Betty will play the rôle of questioner. Her job will be to try to get Arthur to produce arguments for THESIS A, and to reply to those arguments with objections. Arthur's job is then to reply to her objections. She may then furnish a counter-reply to his reply. This back-and-forth process, characteristic of dialectic, may be carried on long enough to satisfy both participants. However, in the present illustration, we will keep the sequences manageably short. Having terminated one sequence, Betty must then ask Arthur if he has another argument for THESIS A. If he does, yet another sequence of objections and replies must be carried out by both participants. When Arthur has run out of arguments for THESIS A, we move to round two, where Betty and Arthur will reverse positions in the game (a dialectical shift). Arthur will be the questioner and Betty the answerer.

In this sample disputation, there will be only two rounds, although in principle several rounds could be allowed. One should try to avoid too much repetition of arguments.

To set up an informative and useful dialectical disputation, it is necessary to start with two theses that are logically opposed to each other – that is, one must be inconsistent with the other – so that not more than one of the theses can be true. Moreover, it is important that the disagreement be a genuine one, and not a mere trifler. To ensure this, it is necessary to start with a concrete situation, one that the participants can mentally picture themselves in, a situation that calls for the necessity of a decision for one thesis or the other. Yet both theses must be sufficiently plausible to win adherents. In other words, there has to be enough room for argument on both sides if the disputation is to be interesting. Moreover, as we saw in Chapter Six, the participants must be reasonably equally matched. In our sample, both factors will be balanced evenly. In principle, however, the two factors could be weighted. That is, the heavy-weight arguer could always be set to argue for the less plausible thesis, and the light-weight arguer could defend the more plausible thesis.

In our illustration, we have set a situation: recently the media reported an airplane crash on a remote mountain in the Andes, where several survivors were stranded for two or three months. Search-and-rescue attempts were eventually abandoned, and the survivors began to slowly starve to death after running out of provisions. Eventually, in order to remain alive, the survivors were forced to consider the possibility of consuming the bodies of their unfortunate fellow passengers who had not survived the crash. Let us suppose that you were a survivor, and you had to make the choice between certain death by starvation and eating the flesh of a human cadaver. What would you do?

Given sufficient imagination, each of us can perhaps have some idea of what we might decide to do. But, more important, what would be the reasons for supporting the decision and for rejecting the alternatives? Here begins the dialectic.

1. ROUND ONE

BETTY: Why is cannibalism always immoral in a civilized society?

ARTHUR: Each of us is part of a larger society that has customs, rules, and codes established for the good of all. One of these customs is the taboo against cannibalism.

BETTY: Established customs are not always above moral criticism. It was the established practice, in Nazi Germany, to send minority groups and political dissenters to death camps. Did that make it right?

ARTHUR: Well, certainly, you can't generalize from one particular society, tribe, nation, or other specific group. I'm talking about the norms and standards set by most civilized nations as majority rules in the history of all civilized countries. Cannibalism has only rarely, if ever, been condoned by the norms of majorities.

BETTY: Still, surely it is one question whether an action is immoral, and another question whether it is a commonly accepted practice. The majority of people have never actually been in a particular situation where cannibalism was necessary in order to survive.

ARTHUR: You don't have to be in that particular situation to know that cannibalism is a harmful practice generally. If it were allowed as a general practice, such a custom would hardly be conducive to civilized behaviour.

BETTY: But how are you deciding what is harmful to society? Don't you have to look at the particular situation? It is one thing to allow cannibalism in a desperate and unusual situation, quite another to recommend it as an every-day menu suggestion.

ARTHUR: Well, I quite agree that there is a difference there. But the problem is that if you do allow it in any situation, then you are approving it in that situation. And who is to draw the line and say that it might be just as acceptable in similar circumstances? Where do you draw the line?

BETTY: That's hard to say, I admit. I think you have to judge each particular situation on its own merits. Do you think we could move on at this point? Do you have another argument for your thesis?

ARTHUR: Yes, here is my second argument. Although cannibalism has, from time to time, been adopted as a general practice by isolated tribes or groups, it has always been condemned by civilized societies.

BETTY: Well, I thought we already dealt with that line of argument. I still maintain that you can't argue something must be right just because some majority supports it.

ARTHUR: Yes, I see your point on that. But what I am now arguing is that we can evaluate certain established practices as better than others. The higher civilized societies have banned cannibalism because its adoption would lead to the breakdown of civilized conduct.

BETTY: Well, I can certainly appreciate the proposition that cannibalism could have harmful consequences if adopted as a general practice. But in the case of someone who would otherwise die by slow starvation, eating the flesh of a cadaver would be life-saving, and hence beneficial. In that situation, it could be a benefit, not a harm.

ARTHUR: Particular judgements of benefits must give way to laws of a civilized society. Murder may be beneficial to all concerned in one particular situation. Perhaps it might even turn out to benefit the victim. That would not make it right.

BETTY: Yes, but there is a difference between murder and cannibalism. Cannibalism need not entail the taking of life.

ARTHUR: Yes, of course there's a difference. But my point is that there

<table>
<tr><td></td><td>could also be a similarity in the two types of cases. Both indicate a disrespect for human life.</td></tr>
<tr><td>BETTY:</td><td>I disagree. The person in the plane crash eats the flesh of someone already dead in order to save his or her own life. It is a life-saving, not a life-destroying, act.</td></tr>
<tr><td>ARTHUR:</td><td>Yes, there is that difference, of course. But I am saying that in both cases if you just stick to individual judgement calls in particular situations, you are ignoring the question of the over-all benefits and norms of adopting this practice in a civilized society.</td></tr>
<tr><td>BETTY:</td><td>But in the airplane crash, we are talking about a crisis situation that is very unusual. The people involved should be allowed to make up their own minds on how to proceed. Their lives are at stake. They are not harming anyone else, so they should be free to choose and follow their own individual consciences.</td></tr>
<tr><td>ARTHUR:</td><td>They have to remember that their act of cannibalism will have an effect on others in their society. In this particular case, there was wide media coverage, and the act of cannibalism made headlines all over the world. Giving approval and sanction to such an act does have an effect on how we see ourselves as a civilized people. The pathetic picture of some poor devil munching on his wife's thigh-bone is hardly conducive to our standards of the dignity of mankind.</td></tr>
<tr><td>BETTY:</td><td>But isn't life more important than dignity?</td></tr>
<tr><td>ARTHUR:</td><td>Not always. Facing death with dignity, I believe, is an extraordinary mark of courage and selflessness, valued by civilizations that keep records of heroic acts.</td></tr>
<tr><td>BETTY:</td><td>Very stirring. But I still think it is presumptuous of those of us not confronted with the particular situation to demand the survivors conform to customs they may not agree with. We still seem to disagree. Do you have any other arguments to put forward?</td></tr>
<tr><td>ARTHUR:</td><td>Just one more, although I have already alluded to it in passing. I feel that if even isolated incidents of cannibalism are allowed or approved, there could be a slippery-slope effect. Such an act might lead to the approval of other malicious practices. All such actions contribute to a demoralization and degeneration of any civilized society.</td></tr>
<tr><td>BETTY:</td><td>Yes, I can appreciate that. But I still don't think that the people in the airplane crash are being uncivilized in eating flesh. Everyone can see that the situation is desperate. For this reason, such a particular incident is not likely to lead to a sudden upswing in human flesh as a popular menu item. As I have said before, there is a clear difference between one isolated incident and any possibility of adoption of cannibalism as a widespread custom.</td></tr>
<tr><td>ARTHUR:</td><td>But that's just the problem. I don't think you can just neatly draw a line and feel sure that people are going to appreciate the difference. Eating human flesh shows disrespect for a human being. Once you allow such disrespect even on a modest basis, it</td></tr>
</table>

indicates a tear in the fabric of human decency.

BETTY: Yes, of course, but this is a bit silly in the case of cannibalism, as opposed to other practices, like murder or theft. There's no need for mass cannibalism. I just can't get too worried about it, given the fact that it seems so unlikely.

ARTHUR: True, but likelihood is not the point. It's the wrongness of it that is the point. Moreover, even though it may be unlikely, that could change. Even now, in some countries, widespread adoption of eating the flesh of the dead could alleviate and even entirely eradicate starvation and malnutrition. I mean it seems bizarre at first. But if you get enough people re-educated to think of human flesh-eating as acceptable, it could solve a lot of problems.

BETTY: Yes, but I am talking about what is likely to happen, given present attitudes and circumstances. You are only talking about what could happen. There is a big difference.

ARTHUR: Granted. But I still maintain that likelihood is not the issue. Even if it were, who can calculate the effects of highly publicized actions on the fabric of the social structure? Some actions could have much more impact on how people think than we realize.

BETTY: I agree, it's hard to calculate such effects. Even so, I think that sometimes a person might be right to commit an act that might tend to demoralize or degrade herself or others. For example, I might be right to tell a secret that would demoralize a lot of people, if it were necessary to save someone's life by doing so.

ARTHUR: Agreed, but I think there is a difference between truth-telling and cannibalism. There may be times when another duty – say the duty to save a life – may over-ride the duty to tell the truth. However, I remain unconvinced that it could ever be my duty to eat the flesh of another human being.

BETTY: Yes, but what's the difference? If you can break one rule in exceptional circumstances, why not another?

ARTHUR: The difference is that the revulsion against cannibalism is severe, because it is so unusual and so thoroughly revolting and debilitating. I don't think it can ever be condoned, even under the most trying circumstances.

BETTY: I still don't see the difference. The fact that it's a rule that won't be broken very often doesn't mean there aren't any circumstances in which we couldn't imagine breaking it. Suppose you only had to swallow one little bit of flesh or the whole world would perish?

ARTHUR: Well, it's hard to know what to say to answer a purely imaginary dilemma like that. It's not a realistic situation. Even so, allowing the whole world to perish might not be a choice that would be compatible with respect for life or for the continuance of civilized society. It is forced choice between two evils. It would not make cannibalism any less demoralizing.

BETTY: Wouldn't you have to say that cannibalism would be the right action in that situation?

ARTHUR: No, I still wouldn't feel it would be right, even if I were forced to

do it to save humanity. In itself, it is still not an acceptable
action for any civilized person. It is no less degrading and
destructive to one's self-respect as a civilized person.

BETTY: Well, I would disagree. Perhaps we could go ahead to round two
so I can have a chance to advance my own arguments against
that.

ARTHUR: Agreed.

2. ROUND TWO

ARTHUR: Well, what is your first argument for THESIS B?

BETTY: Human intelligence, employed in a difficult situation, has been
necessary for the survival and development of civilized
societies. The idea of cannibalism seems revolting, but perhaps
this response is a purely emotional one, based on traditional
fears. However, the airplane crash situation is a novel one, and
the intelligent person must reason the situation out for herself.
The conclusion: in this particular situation, survival is more
important than sticking to traditional customs.

ARTHUR: The requirements of survival could be used to justify any
crimes and excesses. Any criminal could justify his crime if he
could use the argument that it was necessary for him to
survive. It would mean that anything goes.

BETTY: No, I'm not saying that survival is the only factor. It's just that
in the airplane crash situation, the intelligent person must
weigh the values of survival against adherence to tradition. In
this particular case, we have no right to condemn a person who
decides that survival is more important *in this particular
situation*.

ARTHUR: But in a civilized society, survival can only be pushed so far as a
justification for action. If you push this argument too far, you
are talking about a regression to primitive and barbaric
practices, an erosion of the values of civilized societies.

BETTY: Well, I still think it is more civilized for each person to make
the decision based on his or her own free and reasoned choices.
Unreasoned social uniformity is, to my way of thinking, more
the hallmark of a barbaric society.

ARTHUR: We seem to disagree about what "civilized" means. However,
could we push on to another argument for THESIS B? If you have
one.

BETTY: Of course. My second argument is that scientific and technological progress is one aspect of the development of civilized
society. These changes and discoveries require the adaptation
of each new generation to new circumstances. Consequently,
old, traditional beliefs must be given up. Only by intelligent
adaptation to new situations can we keep pace and advance
civilization. The traditional idea that flesh is sacred has to give
way to our newer ideas based on modern medicine and
physiology. There should not be such a dark or mysterious
taboo surrounding dead bodies. At any rate, such a taboo should
not be the determining factor in deciding whether cannibalism
is morally acceptable or not.

ARTHUR: Well, yes. Certainly, tradition must be challenged and questioned. But that does not mean there isn't a certain wisdom in it. You can't just throw traditional values out, the minute there is some difficulty in adhering to them.

BETTY: Well, I agree. But in the present case, I think tradition should be challenged because it may be just based on superstition.

ARTHUR: You think so? Surely the body of a recently dead person deserves our respect. To eat it is revolting, because it indicates a disrespect for the person who once was – or at least was in that body.

BETTY: Well, I agree that bodies should be treated with respect, but I think there are limits to that respect. Perhaps that person might prefer that you should make use of their body to save your life. It's not a matter of disrespect but of necessity.

ARTHUR: Still, the whole idea is revolting and ghoulish. And how can you say that the victim would approve of it if he isn't around to tell you one way or the other? Maybe it might be a violation of his religious beliefs. All various cultures, despite their differences, feel that the body is sacrosanct and deserves respect. Cannibalism violates that respect, and is therefore shunned.

BETTY: Just because society does not approve of a particular action, that does not make it wrong. History has sometimes judged the society to be wrong and the individual right. Are you advocating unthinking conformism?

ARTHUR: No, of course not. I think our belief that flesh-eating is revolting is based on sound cultural values.

BETTY: Well, look at it this way. Undergoing open-heart surgery is a pretty revolting procedure, but it is worthwhile if it saves your life. Just because the process is bloody or unpleasant does not mean that it should not be undergone for a greater end.

ARTHUR: Yes, but the cases are different. Heart surgery is upsetting, but really it is just another form of medical treatment. Feasting on the dead is altogether different. It is revolting and ghoulish, an unjustified invasion of the person's body. A patient has to give consent to surgery, whereas cannibalism is an invasion of a person's right to control his body.

BETTY: But if the person is really dead and cannot possibly ever return to conscious life or awareness, how are his or her rights violated? As you yourself insist, there is a difference between a live person and a dead body.

ARTHUR: Well, I agree, of course. But we still owe respect to the dead body of a person, for it once was a person. The body is still sacred, even if the person is dead.

BETTY: But why, Arthur? Isn't it just superstition at bottom? What reason can you give?

ARTHUR: I believe that the person is a unity made up of soul and body. Even when the two are no longer united, the relic of the body is still worthy of awe and respect.

BETTY: Well, I still don't see why, in the airplane crash, the value of saving life is not the most important factor. I agree that bodies should be respected, but I don't see that we need carry such

257

respect so far as to cause the death of another. I mean people donate their organs in order to save the lives of another. It seems ghoulish to some, but modern medicine has modified our traditional fears – and, I think, rightly so.

ARTHUR: Yes, but an organ donor must give consent. In the airplane crash, the victims could not be consulted. And the eating of human flesh is forbidden by the majority of people today. No wonder, too – if cannibalism were a possibility, wouldn't all of us experience anxiety at the prospect that our flesh might be eaten after we are not around to stop it?

BETTY: But is that likelihood a realistic possibility? Cannibalism is hardly likely to become a popular activity.

ARTHUR: Well, we've already been through that one before. If a tolerance to the idea became widespread, it could be economically feasible as a solution to food shortages. But I am repeating myself. Shall we close off the argument here?

BETTY: Agreed.

3. Commentary

Arthur and Betty have not reached agreement on the correctness of THESIS A or of THESIS B. Neither thesis has been refuted, nor has either one been conclusively established. Still, the dialogue has not been uninformative. There has been positive increment of learning, both for the participants and for the observers.

Through the criticism of the other's objections, each has refined and clarified his position. Furthermore, they do agree on some things, and the particular propositions on which they disagree have been more clearly located. In regard to their basic moral principles, Arthur emerges as a believer in adherence to general rules and principles in the best interests of civilized societies. Betty is more of an individualist; she believes that each person needs to use intelligence to think through each individual case on its own merits. They differ in other ways, too. Arthur puts a high premium on his beliefs about the sanctity of the body. Betty partially agrees, but places more value on the principle of saving the life of an individual.

So it turns out that Betty and Arthur are well and evenly matched as participants, and that their theses are also well matched. One can clearly see that the sequences of objections and replies could have been extended further. It is only by agreement of participants that a particular line of inquiry and argument is terminated.

Of course, when discussing deeply held traditions and fundamental moral principles, the dialectic is not likely to terminate in a speedy resolution in favour of one thesis or the other. Such a discussion is a hard test of any model of argument, but only dialectic stands a chance of even limited gains. Complete resolution of the opposition is almost too much to be hoped for. Still, better understanding of the nature of the disagreement is not to be taken lightly. It can lead to tolerance, understanding, and respect. Thus, dialectic has a better chance of avoiding fallacy, aggression, and warfare than does the debate or the quarrel.

Notice, however, that dialectic demanded obedience to certain conventions and rules. Each participant kept his or her replies short and to the point. Each agreed to terminate the line of argument and start a new one when no further headway seemed to be forthcoming. Each took turns, and at least made an earnest attempt to avoid the more annoying and obvious of the fallacious strategies that could have been employed by a participant. Such co-operation is a requirement for dialectic. Otherwise, a descent to the level of the quarrel or debate threatens.

4. How to Conduct a Disputation

Our sample disputation is a kind of idealized model of what a dialectical interchange should look like. In conducting the discussion group exercise of Chapter Six, the sample disputation can be used as a model of what a final write-up could look like. The particular topic and arguments will be quite different. Indeed, the dialogue you engage in, and the report you write up, may differ in many ways from the one between Arthur and Betty. The important thing is that the rules of dialectic given in Chapter Six be followed in setting up your own disputation.

Note

The topic, and many of the arguments, of this sample disputation were suggested by a group of four students: Ralph Collins, Joseph Berezowski, Tom Fijal, and Cathy Breckman. We are grateful to Professor Peter Miller for suggesting their project as an interesting example of group discussion. The concept of a discussion group report is outlined in B.C. Postow, "Independent Discussion Groups for Introducing Philosophy," *Teaching Philosophy* 1,(1975):51–54.

Appendix II:

Debate on Capital Punishment

This interchange is an excerpt from the House of Commons Debates on the subject of capital punishment. The debate is particularly interesting because several of its arguments appear to involve fallacies we have studied.

MR. BRISCO: Thank you, Mr. Speaker. In view of the fact that Bill C-84 is one of those rarities of the parliamentary process in which we have a free vote, or a so-called free vote, in which we have a mutual sharing of thinking on both sides of the House in regard to both sides of the coin, I consider it incumbent upon me to present to my constituents both sides of the question. However, when my constituents are made aware of the fact that the Solicitor General handed out a contract to the research and systems development branch of his department – although he says that the views of the author do not necessarily represent his own views – and when I see in the report that the author, Mr. Ezzat A. Fattah of Simon Fraser University, an institution of cynics and socialists, tells the Canadian people that it seems that the better educated tend to be more tolerant in their attitudes toward punishment and more opposed to the death penalty than the less educated, and when he goes on to indicate that those who are in poor financial circumstances are more apt to be in favour of capital punishment than the well-to-do, then I am inclined to turn that around.

It is very simple to turn that around, Mr. Speaker, and to ask this question. Is Mr. Fattah suggesting, since the people of Kootenay West, on the basis of the poll that I conducted, support capital punishment by the astonishing figure of 92%, that this is a reflection on their intelligence? Is this, then, a reflection on their income? Is this, then, a reflection on the fact that they are uneducated? Is that what this gentleman is implying? Indeed, the same expression could be applied to every riding in Canada. To extend the argument a little further, those areas of the country that show a high degree of support for capital punishment are, according to this report, those sections of Canada where there are large numbers of unemployed and uneducated people.

I have a great deal more regard for the intelligence of my constituents, the hard working and industrious people of my riding, than apparently this report or this fellow from Simon Fraser University has. He goes on to indicate that there is a very strong pattern in favour of the death penalty associated with income and religious belief. He concludes that the few studies which have attempted to examine some psychological determinants of attitudes to the death penalty tend to show that people who approve the death penalty are people who are insecure, who were severely brought up and who are socially maladjusted. That, Mr. Speaker, is exactly what this professor, so-called, has said. What a condemnation of the Canadian people! How can the minister possibly accept the thrust of this man's report when he makes a statement like that?

MR. ALLMAND: Let us hear the quotation verbatim.

MR. BRISCO: I just gave it to you. He says that the few studies which have attempted to examine some psychological determinants of attitudes to the death penalty tend to show that people who approve of the death penalty are people who are insecure, who were severely brought up and who are maladjusted socially.

MR. ALLMAND: Read the rest.

MR. BRISCO: People who are opposed to it are people who feel secure, who were brought up by mild disciplinary methods and who are pleased with life.

Is that a statement that the minister can accept? Is that the statement that this House accepts? Is that the position of this government?

THE ACTING SPEAKER [MR. TURNER]: Order, please. Solicitor General arises on a point of order.

MR. ALLMAND: Mr. Speaker. I rise on a question of privilege. My department has financed many studies, including one by Professor Bibby of the University of Lethbridge and Professor Grenier, and so on. These many studies have presented many views, some of which I think would support the [honorable] member's position. I just want to make certain that these studies are done from time to time and they do not express the views of the department, as the [honorable] member said at the beginning. Even in that regard, I do not think he was being fair because he quoted at random from the report. He could also quote from other reports financed by our department which give the opposite view such as those expressed by Professor Bibby and Professor Grenier.

MR. BRISCO: Mr. Speaker, perhaps I could respond to the minister. He has indicated that his department has introduced other reports which apparently deal with the other side of the coin. That is the very point I was making a few minutes ago in respect of which the minister challenged me.

MR. ALLMAND: Because you were wrong.

MR. BRISCO: I was wrong; you just finished telling the House –

THE ACTING SPEAKER: Order, please. I would suggest to the Solicitor General and the [honorable] member that they are getting into a debate. I suggest the [honorable] member get back to Bill C-84.

MR. BRISCO: Thank you, Mr. Speaker. It would seem that the Solicitor General is overly concerned about my remarks.

AN [HONORABLE] MEMBER: He is a bit touchy.

MR. BRISCO: Yes, he is a little touchy today. If the minister wishes, I am quite prepared to read into the record the opening statements by this individual, and if he cares to trot over some books and reference papers in support of capital punishment, I would be pleased to read those into the record also.

MR. ALLMAND: You were giving just one side of the story and there are two sides.

AN [HONORABLE] MEMBER: You never circulated the other.

MR. ALLMAND: Oh, yes.

AN [HONORABLE] MEMBER: But not widely.

THE ACTING SPEAKER: Order, please. The [honorable] member for Kootenay West has the floor, and I suggest that we listen to him.

AN [HONORABLE] MEMBER: Tell the minister to be quiet.

MR. BRISCO: It is unfortunate that I am getting all this heckling from the Solicitor General. Quite frankly, I do not mind it too much because I do not find it difficult to respond to.

MR. ALLMAND: I didn't start it.

MR. BRISCO: One of the most strange parts of this capital punishment matter and the hanging of murderers is the absence of concern and alarm by the public when a convicted child murderer or rapist pays the supreme penalty in a prison at the hands of other inmates. In that case do we see agonizing in the press, and do we see many letters to the editor?

MR. ALLMAND: Yes.

MR. BRISCO: Do we then hear the expression of concern in the House, or elsewhere?

MR. ALLMAND: Yes, very often.

MR. BRISCO: The minister knows, as does everyone here, that the degree of concern expressed in respect of justice taken into the hands of inmates in this type of case is very minimal compared with the type of press and concern reflected in the House and outside when such an event takes place as the one in Saskatchewan last year when a murderer took the life of two small children. We know that according to the testimony of the murderer, one small child asked him, after witnessing the death of the other, "Are you going to kill me, too?" When I think of that kind of response from a child, when I think of the brutalizing of children, and when I think about our concern for the victim – which so rarely seems to be expressed – I think it is time people of that ilk started to pay the supreme penalty. I have no hesitation whatever in saying that society simply does not need these mad dogs. Society knows full well that these people are beyond the pale, and 25 years in jail is not going to contribute anything to their recovery.

Today, in cities such as New York we see apartments and office buildings under security, with armed guards and wire fences. The same situation exists in Philadelphia and Chicago. How far are we away from this situation in Canada? How long will it be before we see this kind of security escalate in Vancouver and Montreal? How long will it be before we see security escalate in the House of Commons? In my view, we are progressively moving into jungle warfare, with criminals having declared an open season on innocent victims. The question is: Should we stand idly by, or should we begin to fight? Are we going to allow this guerrilla warfare to overwhelm us, and turn the other cheek?

Notes

Appendix II

1   The debate is recorded in *Hansard*, 14 June 1976, 14464–14465. A useful exercise: analyse those passages where the speaker puts forward an argument that may be fallacious.

Another exercise: look through records of debates, like parliamentary transcripts, and see what fallacies can be found.

Selected Readings

Argument: The Logic of the Fallacies is designed to be used flexibly in introducing elementary level courses. It can be used by itself, in one semester, or combined with other texts in a longer course in various ways. It can be used with a text in formal logic. Or it can be used with any one of the numerous other texts of informal logic. Other informal texts tend to be marked by a lesser emphasis on the theory of the fallacies, but many of them are a bountiful source of examples and specimens for analysis. Accordingly, we begin by listing some texts in formal and informal logic that we think deserve consideration. Then we will give other useful readings on various topics of interest.

1. Formal Logic

Irving M. Copi, *Symbolic Logic*, 5th ed. (New York: Macmillan, 1979).
W.V. Quine, *Methods of Logic*, 3rd ed. (New York: Holt, Rinehart and Winston, 1972).
E.J. Lemmon, *Beginning Logic* (London: Nelson, 1965).
Bas C. van Fraassen, *Formal Semantics and Logic* (New York: Macmillan, 1971).
John Woods, *Proof & Truth* (Toronto: Peter Martin Associates, 1974).

2. Informal Logic

R.H. Johnson and J.A. Blair, *Logical Self-Defense* (Toronto: McGraw-Hill Ryerson, 1977).
W. Ward Fearnside and William B. Holther, *Fallacy: The Counterfeit of Argument* (Englewood Cliffs: Prentice-Hall, 1959).
Michael Scriven, *Reasoning* (New York: McGraw-Hill, 1976).
David Hackett Fisher, *Historians' Fallacies* (New York: Harper & Row, 1970).
John Woods, "What is Informal Logic?" *Informal Logic: The First International Symposium*, ed. J. Anthony Blair and Ralph H. Johnson (Pt. Reyes, California: Edgepress, 1980), 57–69.
J. Anthony Blair and Ralph H. Johnson, "The Recent Development of Informal Logic," *Informal Logic: The First International Symposium*, ed. Johnson and
264

Blair (Pt. Reyes, California: Edgepress, 1980), 3–28.

S. Morris Engel, *With Good Reason* (New York: St. Martin's Press, 1976).

3. General Reference

C.L. Hamblin, *Fallacies* (London: Methuen, 1970). This book is the only thorough and scholarly treatment of the subject of the fallacies as a whole. It is an essential reference tool in this field.

4. Periodicals

The *Informal Logic Newsletter* publishes much useful material, including examples of fallacies and interesting arguments, evaluations of textbooks, course outlines in informal logic, and scholarly articles on informal logic. Other useful journals to consult are the *Journal of Philosophical Logic, Logique et Analyse*, and *Philosophy and Rhetoric*.

5. Ad Baculum and Ad Populum

Dwight Van de Vate, Jr., "The Appeal to Force," *Philosophy and Rhetoric* 8(1975):43–60.

John Woods and Douglas Walton, "Ad Baculum," *Grazer Philosophische Studien* 2(1976):133–140.

Charles Kielkopf, "Relevant Appeals to Force, Pity, and Popular Pieties," *Informal Logic Newsletter* 2(1980):2–5.

Douglas N. Walton, "Why is the *Ad Populum* a Fallacy?" *Philosophy and Rhetoric*, 13, 1980, 264–278.

6. Ad Hominem

Henry W. Johnstone, Jr., "Philosophy and the *Argumentum Ad Hominem*," *Journal of Philosophy* 49(1952):489–498.

E.M. Barth and J.L. Martens, "*Argumentum Ad Hominem:* From Chaos to Formal Dialectic," *Logique et Analyse* 77–78(1977):76–96.

John Woods and Douglas Walton, "Ad Hominem," *The Philosophical Forum* 8(1977):1–20.

7. Debate and Rhetoric

Otto F. Bauer, *Fundamentals of Debate: Theory and Practice* (Glenview, Illinois: Scott Foresman, 1966).

Chaim Perelman and L. Olbrechts-Tyteca, *The New Rhetoric* (Notre Dame and London: University of Notre Dame Press, 1969).

8. Inductive Argument

Brian Skyrms, *An Introduction to Inductive Logic* (Belmont, California: Dickenson, 1966).

Stephen K. Campbell, *Flaws and Fallacies in Statistical Thinking* (Englewood Cliffs: Prentice-Hall, 1974).

Henry E. Kyburg, Jr., *Probability and Inductive Logic* (London: Collier-Macmillan, 1970).

9. Ad Verecundiam

John Woods and Douglas Walton, *"Argumentum* Ad *Verecundiam,"* Philosophy *and Rhetoric* 7(1974):135–153.
Martin Gardner, *Fads and Fallacies in the Name of Science* (New York, Dover, 1952).

10. Plausibility Theory

Nicholas Rescher, *Plausible Reasoning* (Assen-Amsterdam: Van Gorcum, 1976).

11. Dialectical Games

C.L. Hamblin, *Fallacies* (London, Methuen, 1970) Chapter 8.
Nicholas Rescher, *Dialectics* (Albany: State University of New York Press, 1977).
C.L. Hamblin, "Mathematical Modes of Dialogue," *Theoria* 37(1971):130–155.
J.D. Mackenzie, "How to Stop Talking to Tortoises," *Notre Dame Journal of Formal Logic* 20(1979):705–717.

12. Ad Ignorantiam

John Woods and Douglas Walton, "The Fallacy of *Ad Ignorantiam,"* Dialectica 32(1978):87–99.

13. The Logic of Questions

Lennart Aqvist, *A New Approach to the Logical Theory of Interrogatives* (Filosofiska Studier: Uppsala, 1965).
David Harrah, *Communication: A Logical Model* (Cambridge, Massachusetts: MIT Press, 1963).
Arthur N. Prior and Mary Prior, "Erotetic Logic," *Philosophical Review* 64(1955):43–59.
Nuel D. Belnap and Thomas B. Steel, *The Logic of Questions and Answers* (New Haven: Yale University Press, 1976).
Jaakko Hintikka, "The Semantics of Questions and the Questions of Semantics," Amsterdam, North-Holland, *Acta Philosophica Fennica*, 28, 1976.

14. Arguing in a Circle

Oliver Johnson, "Begging the Question," *Dialogue* 6(1967):135–150.
David H. Sanford, "Begging the Question," *Analysis* 32(1972):197–199.
John A. Barker, "The Fallacy of Begging the Question," *Dialogue* 15(1976):241–255.
John Woods and Douglas Walton, "Petitio Principii," *Synthese* 31(1975):107–127.
John Woods and Douglas Walton, *"Petitio* and Relevant Many-Premissed Arguments," *Logique et Analyse* 20(1977):97–110.
John Woods and Douglas Walton, "Arresting Circles in Formal Dialogues," *Journal of Philosophical Logic* 7(1978):73–90.

J.D. Mackenzie, "Question-Begging in Non-Cumulative Systems," *Journal of Philosophical Logic* 8(1978):117–133.
Douglas N. Walton, "*Petitio Principii* and Argument Analysis," *Informal Logic: The First International Symposium*, ed. J. Anthony Blair and Ralph H. Johnson (Pt. Reyes, California: Edgepress, 1980), 38–54.
John Woods and Douglas Walton, "Question-Begging and Cumulativeness in Dialectical Games," *Noûs* to appear in 1981.

15. Equivocation

John Woods and Douglas Walton, "Equivocation and Practical Logic," *Ratio* 21(1979):31–43.
Christopher Kirwan, "Aristotle and the So-Called Fallacy of Equivocation," *Philosophical Quarterly* 29(1979):35–46.

16. Relatedness Logic

Richard L. Epstein, "Relatedness and Implication," *Philosophical Studies* 36(1979):137–173.
Douglas N. Walton, "Philosophical Basis of Relatedness Logic," *Philosophical Studies* 36(1979):115–136.

17. Economic Reasoning

Leonard Silk, *Contemporary Economics: Principles and Issues*, 2nd ed. (New York: McGraw-Hill, 1975).
Kenneth J. Arrow, *Social Choice and Individual Values*, 2nd ed. (New York: Wiley, 1964).
Günter Menges, *Economic Decision Making: Basic Concepts and Models* (London: Longman, 1974).

18. Decision Theory and Games Theory

J.D. Williams, *The Compleat Strategyst* (New York: McGraw-Hill, 1966).
Richard C. Jeffrey, *The Logic of Decision* (New York: McGraw-Hill, 1965).
W.F. Lucas, "Game Theory," *Encyclopedia of Computer Science and Technology*, vol. 8, ed. A.G. Holzman and A. Kent (New York: Marcel Dekker, 1977), 363–392.

19. Composition and Division

John Woods and Douglas Walton, "Composition and Division," *Studia Logica* 36(1977):381–406.
Tyler Burge, "A Theory of Aggregates," *Noûs* 11(1977):97–118.

20. Modal Logic

G.E. Hughes and M.J. Cresswell, *An Introduction to Modal Logic* (London: Methuen, 1968).

21. Epistemic Logic

Richard L. Purtill, *Logic for Philosophers* (New York: Harper & Row, 1971).
Jaakko Hintikka, *Knowledge and Belief* (Ithaca: Cornell University Press, 1962).

22. Varia of Interest

F. Cizek, "The Problem of Fallacies," *Teorie A Metoda* 6(1974):101–116.
John Woods and Douglas Walton, "Post Hoc, Ergo Propter Hoc," *Review of Metaphysics* 30(1977):569–593.
N.R. Hanson, "The Genetic Fallacy Revisited," *American Philosophical Quarterly* 4(1967):101–113.
J. Pashman, "Is the Genetic Fallacy a Fallacy?", *Southern Journal of Philosophy* 8(1970):57–62.
Max Black, *Margins of Precision* (Ithaca and London: Cornell University Press, 1970).
Stephen E. Weiss, "The Sorites Fallacy: What Difference Does a Peanut Make?", *Synthese* 33(1976):253–272.
Richmond Campbell, "The Sorites Paradox," *Philosophical Studies* (1974):175–191.

23. Historical

William and Martha Kneale, *The Development of Logic* (Oxford: The Clarendon Press, 1962).
Augustus de Morgan, *Formal Logic* (London: Taylor and Walton, 1847).
Aristotle, *Topica et Sophistici Elenchi*, translated by W.A. Pickard-Cambridge, ed. W.D. Ross (Oxford: The Clarendon Press, 1958).
Richard Whately, *Elements of Logic* (New York: William Jackson, 1836).
Alfred Sidgwick, *Fallacies* (New York: D. Appleton & Co., 1884).
L.M. de Rijk, "Some Thirteenth Century Tracts on the Game of Obligation," *Vivarium* 14(1976):26–49.
Mary Brown, "The Role of the *Tractatus de Obligationibus* in Mediaeval Logic," *Franciscan Studies* 26(1966):26–35.

INDEX of NAMES

GENERAL INDEX

270